SpringerBriefs in Pharmaceutical
Science & Drug Development

For further volumes: http://www.springer.com/series/10224

Thomas Catalano

Good Laboratory Practices for Forensic Chemistry

 Springer

Thomas Catalano
PharmChem Analytical Consultants, LLC
Buffalo Grove, IL, USA

ISSN 1864-8118   ISSN 1864-8126 (electronic)
ISBN 978-3-319-09724-4   ISBN 978-3-319-09725-1 (eBook)
DOI 10.1007/978-3-319-09725-1
Springer Cham Heidelberg New York Dordrecht London

Library of Congress Control Number: 2014947528

Printed on acid-free paper

Springer is part of Springer Science+Business Media (www.springer.com)

*This book is dedicated to my wife, Jeanie,
who helped with the manuscript
and is a lifelong friend.*

Contents

Abbreviations

Accuracy	The closeness to the true value
ASCLD	American Society of Crime Laboratory Directors
Between variance	The variance between two data sets
CI	Confidence interval
CFR	Code of Federal Regulations
Column capacity	The columns ability to retain an analyte
Correlation coefficient	The degree of the relation between two variables
Daubert	A rule of evidence introduced by the Supreme Court
Diode array	UV detector measuring many wavelengths at the same time
DNA	Deoxyribose nucleic acid
DOF	Degrees of freedom
DOJ	Department of Justice
DSC	Differential scanning calorimetry
Expanded uncertainty	The reporting value of uncertainty
FDA	Food and Drug Administration
Federal register	Bell Shape curve resulting from plotting the data frequency vs. data spread from physical entity and can be specifically describe by a mathematical equation.
Flow rate	The rate of fluid moving through a column (mL/min)
F-statistic	The statistic for comparing variances
Fyre	A rule of evidence introduced by the Appelate court
GC	Gas chromatography
GLP	Good laboratory practices
GMP	Good manufacturing practices
Gradient	A change of physical properties and a given rate (temperature, composition, concentration, etc.)
Headspace	The vapor space above a liquid or a solid
HPLC	High performance liquid chromatography
Hypothesis	A theoretical proposal
ICH	International Committee for Harmonization
Intermediate precision	The precision

Ionic strength	Concentration of ionic charge in a solution
IQ	Installation qualification
ISO	International Organization of Standards
Isocratic	The physical properties of a system in a constant state
Linear regression	It is the best mathematical fit for a straight line
Linearity	The straight line relationship between two variables
LOD	Limit of detection, the lowest level that can be detected in an analytical procedure
LOQ	Limit of quantitation, the lowest level that can be quantitated in an analytical procedure
Median	The center value of a set of data
Method transfer	The transfer of a procedure to another location
MS	Mass spectrometry
NAS	National Academy of Science
NCFS	National Commission on Forensic Science
NIST	National Institute of Standards and Technology
Normal distribution	A federal agency which maintains, controls, and houses all federal documents
OQ	Operational qualification
Outlier	A value that is statistically different than the rest of the data set
pH	A measure of the hydrogen ion concentration in a solution
PQ	Performance qualification
Precision	The measure of closeness of a set of data
QA	Quality assurance
QC	Quality control
QRM	Quality risk management
Reference standard	Material which is accurately characterized
Repeatability	The within run precision of a data set
Residual	The difference between a theoretical (observed) value and a calculated value
Robustness	The evaluation of the variation of critical parameters on the impact on the procedure against established criteria
RRT	Relative retention time
RT	Retention time
Sample replicate	Multiple individually prepared samples
Scientific method	The procedure that follows a specific approach for the evaluation of natural phenomenon
Significance	Statistical approach that allows to determine if two data sets come from the same or different populations
SOP	Standard operating procedure
Specificity	The ability to distinguish multiple entities from each other within a process
Standard deviation	The distribution of a data set about the mean
Standard error of the mean	Represents the variation among the means

Standard uncertainty	Combined uncertainty after all mathematical operations are completed (addition, subtraction, multiplication, division)
SWG	Scientific Working Group
System suitability	Set of critical parameters that must be met for the system to be acceptable for use
t-Statistic	The statistic that is used for evaluating confidence and significance about the mean
USP/NF	The United States Pharmacopeia/National Formulary
Viscosity	The potential of a solution to flow

About the Author

Tom Catalano obtained his Ph.D. in pharmaceutical sciences at St. John's University in Queens, New York. He started his forensic career as a member of the New York City Police Department (NYPD) Crime Laboratory doing testing in various areas including drugs, gunshot residue, arson investigation, blood alcohol determination, glass analysis, soil comparison, inks and dyes analysis, paint chip analysis, and explosive residue. Dr. Catalano published seven articles on various subjects while a member of the NYPD Crime Laboratory. He later moved on to a career within the pharmaceutical industry, where he has held various positions—including the Global Sr. Director of Analytical R&D—with major pharmaceutical companies such as G.D. Searle, Pharmacia, Pfizer, and Theravance, giving him vast experience in providing analytical support to dosage forms, drug substances, and biologicals. Dr. Catalano has made significant contributions to the development and registration of many products that are currently on the market. He is not only an expert in analytical chemistry but also in the FDA Guidances, guidelines, and work processes that are applicable to forensic chemistry. Currently, Dr. Catalano is president of PharmChem Analytical Consultants LLC., a consulting company that provides service to the pharmaceutical industry, food industry, and forensic chemistry.

Chapter 1
Introduction

Any endeavor which reaches its pinnacle is considered an art. The genius required to produce a musical symphony is not fundamentally different than that needed to develop and introduce a new fundamental concept of nature, such as Einstein's *Space and Time Relativity*. Both require going beyond normal rational thinking and having courage to bring that thinking forward. Science advancement is based on small increments of gained knowledge, which is then extrapolated towards a much larger concept, which may not be totally explainable by the current data. Forensic analysis lies between the boundaries of art and science, in that it often requires an intuitive approach since facts are often in short supply and analytical data is not ideal. However, since we are aware of these limitations, the introduction of rigorous guidances and guidelines can help control our natural tendencies to fill in the gaps. Our goal is to maintain scientific objectivity, however regardless whether we are evaluating fingerprints or chemical spectra the results are not free from human subjectivity. And that's the way it should be. No replicate analysis, even from the same sample, is expected to be identical; it is up to the knowledge and experience of the analyst to determine whether the differences are significant enough to determine that the analysis indicates a difference or no difference. However, there are branches of science such as, statistics and probability that can aid the analyst in evaluating the variability, confidence, and uncertainty associated with analysis method. Data obtained within method variance, confidence limits, and uncertainty value, although different, will be considered indistinguishable. Although the analyst has the final responsibility for the results, disregarding the methods variance, confidence limits, and uncertainty value will require a detailed justification which would be acceptable to his/her peers.

The Pharmaceutical Industry is a highly regulated industry and has developed a rigorous set of guidances, guidelines, and work process, usually as a result of major mistakes and oversight, whose concepts could be evaluated for application to the forensic chemistry discipline. The analysis of evidence and the presentation of the results in a court of law are analogous to the analysis of a drug product and the

© Thomas Catalano 2014
T. Catalano, *Good Laboratory Practices for Forensic Chemistry*,
SpringerBriefs in Pharmaceutical Science & Drug Development,
DOI 10.1007/978-3-319-09725-1_1

submission of the data to a regulatory agency for permission to market the product. Both are responsible for the public safety and the life and death of individuals.

This book will describe in detail the relevant guidances, guidelines, and work processes which may be revised to meet the application to forensic chemistry without changing the conceptual rigor initially intended in the document. Areas such as Good Manufacturing Practices (GMP), Good Laboratory Practices (GLP), International Committee on Harmonization (ICH), Quality Assurance (QA), Quality Risk Management (QRM), Documentation, Reports, to name a few will be reviewed and considered for application to forensic chemistry.

Chapter 2
What Is Science

In order to consider any action to be scientific, that action should follow the concept of the *Scientific Method*. The word science comes from the Latin "scientia," meaning knowledge [1, 2, 4].

Science is defined in accordance to Webster's New Collegiate Dictionary, science is "knowledge attained through study or practice," or "knowledge covering general truths of the operation of general laws, especially as obtained and tested through the scientific method and concerned with the physical world" [4].

Therefore, science refers to a system of acquiring knowledge. This system uses observation and experimentation to describe and explain natural phenomena. The term science also refers to the organized body of knowledge people have gained using that system. What is the purpose of science? Perhaps the most general description is that the purpose of science is to produce useful models of reality.

Most scientific investigations use some form of the *Scientific Method*. Science can be referred to as pure science to differentiate it from applied science, which is the application of research to human needs. Fields of science are commonly classified along two major lines:

- Natural sciences, the study of the natural world
- Social sciences, the systematic study of human behavior and society

2.1 The Scientific Method [3]

The scientific method is a logical and rational order of steps by which scientists come to conclusions about the world around them. The scientific method helps to organize thoughts and procedures so that scientists can be confident in the answers

© Thomas Catalano 2014
T. Catalano, *Good Laboratory Practices for Forensic Chemistry*,
SpringerBriefs in Pharmaceutical Science & Drug Development,
DOI 10.1007/978-3-319-09725-1_2

they find. Scientists use observations, hypotheses, and deductions to make these conclusions. The steps of the scientific method are:

Observation/Research
Hypothesis
Prediction
Experimentation
Conclusion

The observation is done first so that you know how you want to go about your research. The hypothesis is the answer you think you'll find. The prediction is your specific belief about the scientific idea: The experiment is the tool that you create to answer the question, and the conclusion is the answer that the experiment gives.

2.1.1 Observation

This step could also be called "research." It is the first stage in understanding the problem you have chosen. After you decide on what the specific question is, you will need to research everything that you can find about the problem. You can collect information on your science question through books, journal articles, the Internet, or even smaller "unofficial" experiments. For this stage of the scientific method, it's important to use as many sources as you can find. The more the information you have on your science topic, the better the design of your experiment is going to be.

2.1.2 Hypothesis

The next stage of the scientific method is known as the "hypothesis." This word basically means "a possible solution to a problem, based on knowledge and research." The hypothesis is a simple statement that defines what you think the outcome of your experiment will be.

The first stage of the scientific method, the observation, or research stage is designed to help you express a problem in a single question and propose an answer to the question based on what you know. The experiment that you will design is done to test the hypothesis.

2.1.3 Prediction

The hypothesis is your general statement of how you think the scientific phenomenon in question works. Your prediction lets you get specific on how you will demonstrate that your hypothesis is true? The experiment that you will design is done to test the prediction.

An important thing to remember during this stage of the scientific method is that once you develop a hypothesis and a prediction, you shouldn't change it, even if the results of your experiment show that you were wrong.

An incorrect prediction does NOT mean that you "failed." It just means that the experiment brought some new facts to light that maybe you hadn't thought about before.

2.1.4 Experimentation

This is the part of the scientific method that tests your hypothesis. An experiment is a tool that you design to find out if your ideas about your topic are right or wrong.

It is absolutely necessary to design an experiment that will accurately test your hypothesis. The experiment is the most important part of the scientific method. It's the logical process that lets scientists learn about the world.

2.1.5 Conclusion

The final step in the scientific method is the conclusion. This is a summary of the experiment's results, and how those results match up to your hypothesis.

You have two options for your conclusions: based on your results, either you CAN REJECT the hypothesis, or you CAN ACCEPT the hypothesis.

This is an important point. You cannot PROVE the hypothesis with a single experiment, because there is a chance that you made an error somewhere along the way. What you can say is that your results SUPPORT the original hypothesis.

If your original hypothesis didn't match up with the final results of your experiment, don't change the hypothesis. Instead, try to explain what might have been wrong with your original hypothesis. What information did you not have originally that caused you to be wrong in your prediction? What are the reasons that the hypothesis and experimental results didn't match up?

Remember, the experiment isn't a failure if it proves your hypothesis wrong or if your prediction isn't accurate. A science experiment is only a failure if its design is flawed. A flawed experiment is one that doesn't keep its variables under control, and doesn't sufficiently answer the question that you asked of it.

2.1.6 Data

Your data collection must be scientific and professional. Be sure to use a journal to record data from the experiment. This demonstrates organization. Did you repeat the experiment? Repetition lends much more reliability to your data. Repeat it if you can.

2.1.7 Interpretation

The use of tables and graphs are helpful in understanding the data. Ensure that enough data was collected to reach a reliable conclusion. Make sure that you are confident in your final numbers. Science is all about proof.

2.2   Science vs. Art [3]

The arts and sciences are connected. Both the arts and the sciences are not merely connected but evolve from the same human desire. That is our attempt to develop an understanding of the universe, and our attempt to influence things in the universe that are both internal and external to ourselves. The arts and sciences are outcome of human creativity driven by the curiosity by us to develop an understanding of the world around us. Although some argue that art and science are basically the same and are indistinguishable, there are some simple statements that can show practical differences.

Science is about fact…until it's no longer a fact.

Art is about arguing meanings, feelings, and contesting views.

Science is about understanding the world, what's in it, what's beyond it.

Art is about searching within, expanding the world, and determining perceptions.

Science is about natural order.

Art is about justification of thought.

Science is about discovering significance.

Art is about *giving significance*.

Science is objective.

Art is subjective.

Art needs no proof, it cannot be proved. On the other hand, *Science* is based upon theories and hypothesis, which must be proven.

References

1. National Research Council (2009) Strengthening forensic science in the United States, a path forward. National Academies Press, Washington, DC
2. Inman K, Rudin N (2001) Principles and practice of criminalistics. CRC, Boca Raton
3. Harris DA (2012) Failed evidence, why law enforcement resists science. New York University Press, New York, NY
4. Webster's New Collegiate Dictionary (1981), G. & C. Merriam: Springfield

Chapter 3
Forensic Chemistry

3.1 Current State [1, 3, 5]

Over the last two decades, forensic science received very high notability in the public eye. Mainly because of the popularity of television shows such as CSI Miami and CSI New York, to name a few, put a twist on standard police procedure. In this new world of policing, crimes are solved using high tech scientific technologies, very rapidly and with 100 % certainty. However, this common view of modern police work using high tech science as the way of the future in crime solving turns out to look much different in eyes of police and prosecutorial agencies. Over the last two decades, many advances have been made in forensic science such as, DNA technology which has demonstrated that some disciplines in forensic science have made progress in the support of law enforcement. However, there are great differences in the practice of forensic science across various jurisdictions, many due to funding, equipment, and the availability of skilled and well-trained personnel. This fragmentation exists because many of the operational principles and procedures in the forensic science disciplines are not standardized. Generally there are no standard protocols directing the practices for a given discipline. Therefore the quality of practices, in most disciplines, varies greatly because of the lack of adherence to standardized protocols, stringent performance standards, and effective oversight. Because the forensic data is generally used in a court of law (criminal and civil), it is critical to determine whether the forensic data can be accepted as evidence. There are two important questions that must be answered before the court should accept and rely on forensic data as evidence in a court trial. They are (1) the insurance that the forensic discipline and practice is founded on a reliable scientific methodology which has the capacity to accurately analyze evidence with known variation, and level of uncertainty to report its findings and (2) the extent in which the results are based on human subjective interpretation which can introduce bias due to the lack of sound operating procedures and stringent performance standards [2].

© Thomas Catalano 2014
T. Catalano, *Good Laboratory Practices for Forensic Chemistry*,
SpringerBriefs in Pharmaceutical Science & Drug Development,
DOI 10.1007/978-3-319-09725-1_3

With the exception of DNA, our police and prosecutorial agencies are very skeptical about the use of science on how evidence is collected, tested, and how conclusions are drawn from it. In order to try and understand the resistance observed to the use of science in criminal investigation, we should first look at the DNA story. DNA did not develop from within a police-driven forensic investigation, but from a typical scientific approach. Because of the use of the Scientific Method, DNA testing utilizes proven standard protocols, and calculated parameters such as accuracy, precision, selectivity, confidence levels, and robustness based on rigorously analyzed data. Disciplines such as forensic chemistry and toxicology are also derived from the Scientific Method approach, however they are missing the use of a standardized protocol to apply the technology in a manner which would be acceptable in the scientific community which developed and utilizes the technology.

3.2   How to Improve [1, 5, 6]

In order to improve the validity of forensic evidence presented in court, it became part of our law that the judge would act as the gatekeeper for determining which scientific forensic evidence are appropriate for consideration to be presented in court. Since, for the most part, judges are not trained scientist; it was very difficult for judges to decide on the validity of the scientific evidence. Two court rulings were passed over the years to set some criteria for determining whether the scientific forensic evidence was suitable for presentation in court. Frye vs. United States, Court of Appeals of District of Columbia, in 1923, proposed that the scientific approach needed to be sufficiently established so that it had gained general acceptance in the relevant scientific community. This ruling being very general did not contain the specifics in terms how the scientific data was generated and interpreted. Thereby still leaving an open question on the validity of scientific evidence. In 1993 Jason Daubert and Eric Schuller were two children who claimed that they were born with serious defects based on their mother's use of Benedectin during her pregnancy. They sued Merrell Dow the manufacture of the drug. The company submitted an affidavit from a well-credentialed epidemiologist stated he reviewed all the literature involving over 130,000 patients and that no study had found the drug to be capable of causing birth defects. The plaintiffs submitted the testimony of well-credentialed expert who claimed that the drug is capable of producing birth defects. Their conclusions were based on animal studies, pharmacological studies, and some reanalysis of the prior published studies. The judge applied the Frye test rule and deemed the testimony of the plaintiff's experts not admissible because the work done was unpublished and not subject to the normal peer review process and that the data generated was solely for the use in this litigation, thereby ruling in favor of the defendant. Upon appeal, the Supreme Court ruled that current Federal Rules of Evidence superseded and replaced the Frye test of general acceptance test. The rule stated that any scientific, technical, or other specialized knowledge that will assist in the understanding of the evidence or to determine a fact at issue, a witness qualified as an expert may testify in the form of an opinion. However, under the rules the trial judge must see that any and all scientific evidence is not only relevant but reliable. The court said that the methodology must be scientifically valid and that the methodology can be properly applied to facts in issue. The court suggested that several factors should be considered, such as whether the methodology can be scientifically supported, an error rate be determined and has the methodology been published in peer-reviewed professional journals, although publication is not an absolute requirement. Today the Daubert decision consists of methodology to be scientifically supported and contains associated error measurements along with the determination of confidence intervals where appropriate. Although the rules like Frye and Daubert have been identified, they have not resulted in any meaningful limitations on the admissibility of forensic evidence and as a result are not practiced in many state and local jurisdictions. The courts still rely on precedent and every ill-informed decision becomes a precedent binding on future cases.

The courts will not be able to move beyond this misguided precedent approach until real science is brought to bear in assessing the validity and reliability of forensic disciplines and in establishing quantifiable measures of uncertainty in the conclusions of forensic analyses. Recently other groups have immerged to address the problems associated with the validity of the forensic evidence presented in court [4]. One of these groups is the Scientific Working Group (SWG) for various forensic disciplines. The group addresses issues such as Education and Training, Methods of Analysis, Method Validation, Sampling, Documentation, Uncertainty Measurement, and Quality Assurance. The American Society of Crime Laboratory Directors (ASCLD) is another organization working on setting standards for forensic laboratories. Their major concern is the accreditation of the forensic laboratories; however, they are also working on setting standards for Laboratory Management, Code of Ethics, Uncertainty Measurement, Proficiency Testing, Calibration of Instrumentation, Emerging Technology, and active QA/QC programs.

In January 2014, as a result of the National Academy of Sciences (NAS) published report of February 18, 2009, the National Institute of Standards and Technology (NIST) and the Department of Justice (DOJ) created the National Commission on Forensic Science (NCFS) that will work to improve the practice of forensic science by developing guidance and policy recommendations for the US Attorney General. Under this administration, a number of interdisciplinary working groups have been launched to produce technical publications and other forms of critical guidance for the forensic science community. The areas being addressed are Forensic Research, Development of Standards, Guidelines and Best Practices, Scientific Capacity, New Technology and Tools, Workshops and symposia, Education and Training, and International Collaborations. Although there is a recent thrust towards the improvement in the application of forensic science by the creation of a new federal agency NCFS. It is the focus of this book to investigate selective FDA Regulations, Guidances, Guidelines, and Work Processes and generally apply them to the application of forensic chemistry.

References

1. National Research Council (2009) Strengthening forensic science in the United States, a path forward. National Academies Press, Washington, DC
2. Inman K, Rudin N (2001) Principles and practice of criminalistics. CRC, Boca Raton
3. Harris DA (2012) Failed evidence, why law enforcement resists science. New York University Press, New York
4. St Clair JJ (2002) Crime laboratory management. Academic, London
5. Shelton DE (2011) Forensic science in court challenges in the twenty-first century. Rowman & Littlefield, Lanham
6. Hadley K, Fereday MJ (2008) Ensuring competent performance in forensic science. CRC Press, Boca Raton

Chapter 4
Code of Federal Regulations (21CFR) Guidances

4.1 Good Manufacturing Practices (GMP) [1, 3, 4, 7]

As required by law, the Food and Drug Administration publishes regulations in the *Federal Register*, the federal government's official publication for notifying the public of many kinds of agency actions. Federal regulations are either required or authorized by statute. Some address a specific problem or known health hazard, while others, like citizen petition regulations, are administrative or procedural. The rulemaking procedures that we follow come from the US law, Executive Orders (EOs) and memoranda issued by the President, and FDA's own regulations. Title 21 of the CFR is reserved for rules of the Food and Drug Administration. This database contains content that is current as of April 1, 2013. In order for industry to comply with these regulations, standard work processes where developed and are included in the discussions below where appropriate.

4.1.1 Method Development [1]

Methods developed for intended use in forensic chemistry must ensure identity and purity of the material being analyzed. Data from the method developed must meet standards of accuracy and reliability. An analytical method is developed to test a defined characteristic of the material against preestablished acceptance criteria for that characteristic. Early in the development of a new analytical procedure, the choice of analytical instrumentation and methodology should be selected based on the intended purpose and scope of the analytical method. Parameters that should be addressed during method development are specificity, linearity, limits of detection (LOD) and quantitation limits (LOQ), range, accuracy, and precision. Analytical methods are initially developed based on a combination of mechanistic

© Thomas Catalano 2014
T. Catalano, *Good Laboratory Practices for Forensic Chemistry*,
SpringerBriefs in Pharmaceutical Science & Drug Development,
DOI 10.1007/978-3-319-09725-1_4

understanding of the basic methodology and prior experience. Experimental data from early experiments or from existing methodology should be utilized to guide further development. To fully understand the effect of changes in method parameters on an analytical method, you should adopt a systematic approach for a method robustness study (e.g., a design of experiments with method parameters). You should utilize multivariate experiments to understand factorial parameter effects on method performance. Knowledge gained during these studies on the sources of method variation will allow you understand the method performance and determine the critical parameters need to be controlled during the implementation of the method. The analytical method should be written in sufficient detail to allow a competent analyst to reproduce the necessary conditions and obtain results within the proposed acceptance criteria. You should also describe aspects of the analytical procedures that require special attention. The following is a list of essential information you should include in an analytical method:

4.1.1.1   Principle/Scope

A description of the basic principles of the analytical technology (separation, detection, etc.) sample(s) type.

4.1.1.2   Apparatus/Equipment

All required qualified equipment and components (e.g., instrument type, detector, column type, dimensions, and alternative column, filter type).

4.1.1.3   Operating Parameters

Qualified optimal settings and ranges (allowed adjustments) critical to the analysis (e.g., flow rate, components temperatures, run time, detector settings, gradient, head space sampler), integration parameters used for data acquisition.

4.1.1.4   Reagents/Standards

The following should be listed:

- Grade of chemical (e.g., USP/NF, American Chemical Society, High Performance Liquid Chromatography, or Gas Chromatography and preservative free).
- Source (e.g., USP reference standard or qualified in-house reference material).
- State (e.g., dried, undried) and concentration.
- Standard potencies (purity correction factors).
- Storage controls.
- Directions for safe use (as per current Safety Data Sheet).
- Validated or useable shelf life.

4.1.1.5   Sample Preparation

Sample procedures (e.g., extraction method, dilution or concentration, desalting procedures and mixing by sonication, shaking) used for the preparations of individual sample tests. Single preparation is utilized for qualitative and replicate preparations for quantitative tests and information on stability of solutions and storage conditions.

4.1.1.6   Standards Control Solution Preparation

Procedures for the preparation and use of all standard and control solutions with appropriate units of concentration and information on stability of standards and storage conditions, including calibration standards, internal standards, and system suitability standards.

4.1.1.7   Procedure

A step-by-step description of the method (e.g., equilibration times, scan/injection sequence with blanks, samples, controls, sensitivity solution, standards to maintain validity of the system suitability during the span of analysis, and allowable operating ranges and adjustments if applicable.

4.1.1.8   System Suitability

System suitability is an essential test that will respond to the question of reliability of the scientific method each time it is performed. The parameters chosen for evaluation in the system suitability test results from the data generated from the robustness testing of the method and the determination of critical parameters which must be controlled to ensure that the system (equipment, electronics, and analytical operations and controls to be analyzed) will function correctly as an integrated system at the time of use. Examples of parameters which may be utilized for system suitability are listed below:

- Resolution
- Theoretical Plates (for HPLC)
- Tailing Factor
- Peak Asymmetry
- Standard Checks
- Limits of Detection
- Limits of Quantitation
- System Noise
- System Base Line Drift
- Injection Precision (for HPLC)
- Retention Time Reproducibility (for HPLC)
- Recovery Criteria

4.1.1.9   Calculations

The integration method and representative calculation formulas for data analysis (standards, controls, and samples) should be described. This includes a description of any mathematical transformations or formulas used in data analysis, along with a scientific justification for any correction factors used.

4.1.1.10   Data Reporting

A presentation of numeric data that is consistent with instrumental capabilities and acceptance criteria. The method should indicate what format to use to report results (e.g., percentage, weight/weight, and weight/volume, mg/ml, ppm) with the specific number of significant figures needed. For chromatographic methods, you should include retention times (RTs) for identification with reference standard comparison basis, relative retention times (RRTs) for other components detected, and acceptable ranges and sample results reporting criteria.

4.1.1.11   HPLC Method Development [3]

Based on the regulations stated above, an Industry Standard work process has been developed to be consistent with the regulations

Since HPLC is one of the most popular analytical technologies utilized in forensic chemistry, a systematic approach for method development is being described. Before starting method development, the following items should be considered:

- What is the intended use for the method?
- Gather any existing information.
- Gather samples required for method development.
- Establish method criteria.

Examples of existing information which should be obtained are as follows:

- Chemical structure
- Physiochemical properties
- Literature/References
- Related methods

Acquiring the appropriate samples is crucial to the success of developing an acceptable method. Examples of samples to acquire if possible are as follows:

- Reference standards
- Authentic material
- Delivery device
- Capsule shell
- Excipients

The mobile phase is a critical component of an HPLC method and there are many properties that should consider such as:

- Solvent UV cutoff
- pH
- Buffer UV cutoff
- Ionic strength
- Ion-pairing reagent
- Viscosity
- Column compatibility
- Compatibility with mass spectroscopy

Isocratic HPLC Method Development Process

The most frequent type of HPLC method required to develop is an isocratic method. The process for developing an isocratic HPLC method is described below:

- Obtain appropriate samples
- Select columns for evaluation (e.g., C18, Phenyl, CN)
- Set up an initial ACN/buffer gradient

  - 5–95 %, 20 min, 10 min hold

- Observe if any peaks have retention time at the hold time

  - If any peaks observed during hold time, revise gradient for each column evaluated until no peak retention times are found in the hold time

- Choose column that gives greatest number of peaks and best selectivity
- Determine whether an isocratic method is feasible for development

  - Range of all peak retention times in gradient run must be ≤ 40 % of the total gradient time

- Calculate the isocratic solvent strength for the main peak in the ACN gradient
- Run the isocratic condition and determine the k' for first and last peak
- Adjust the isocratic solvent strength to give a k' of approximately 2–20 for first and last peak, respectively
- Calculate equivalent solvent strength for MeOH and THF using the Solvent Strength Conversion Chart shown below in Table 4.1
- Construct a Ten Experiment Solvent Mixture Design Triangle based on the equivalent solvent strengths at the corners obtained from the Solvent Strength Chart
- The sides of the triangle consist of 66 and 33 % of the solvent corners making up that side. The middle is 33 % of each corner of the triangle as shown in Fig. 4.1 below.
- Run each experiment in mixture design triangle

Table 4.1 Solvent strength conversion chart (reverse phase)

MeOH (%)	ACN (%)	THF (%)
5	5	3
10	10	7
15	15	10
20	19	14
25	24	17
30	29	21
35	34	24
40	39	27
45	44	31
50	48	34
55	53	38
60	58	41
65	63	44
70	68	48
75	73	51
80	77	55
85	83	58
90	87	61
95	92	65
100	97	68

[a]Reprinted from [3] permission Springer Science+Business

Exp #1
10.0% MeOH
0.0% ACN
0.0% THF
90.0% Buffer

Exp #2
6.7% MeOH
2.3% ACN
0.0% THF
91.0% Buffer

Exp #3
6.7% MeOH
0.0% ACN
1.7% THF
91.7% Buffer

Exp #4
3.3% MeOH
4.7% ACN
0.0% THF
92.0% Buffer

Exp #5
3.3% MeOH
2.3% ACN
1.7% THF
92.7% Buffer

Exp #6
3.3% MeOH
0.0% ACN
3.3% THF
93.3% Buffer

Exp #7
0.0% MeOH
7.0% ACN
0.0% THF
93.0% Buffer

Exp #8
0.0% MeOH
4.7% ACN
1.7% THF
93.7% Buffer

Exp #9
0.0% MeOH
2.3% ACN
3.3% THF
94.3% Buffer

Exp #10
0.0% MeOH
0.0% ACN
5.0% THF
95.0% Buffer

Fig. 4.1 Example of ten experiment solvent triangle. *Reprinted from [3] permission Springer Science+Business

Interpretation of the data from the ten mixture design experiments

- Identify which experiments give you the greatest number of peaks and best selectivity
- Also observe if any combination of solvents enhance selectivity
- Make minor adjustment between the best identified conditions
- Select the optimum condition for your mobile phase

Gradient HPLC Method Development Process

- Obtain appropriate samples
- Select columns for evaluation (e.g., C18, Phenyl, CN)
- Set up an initial ACN/buffer gradient

  - 5–95 %, 20 min, 10 min hold

- Observe if any peaks have retention time at the hold time.
- If any peaks observed during hold time, revise gradient for each column evaluated until no peak retention times are found in the hold time.
- Run a methanol gradient of equivalent solvent strength to the ACN gradient.
- Choose column that gives the greatest number of peaks and best selectivity.
- Choose which of the solvent gradient (ACN or MeOH) that displays greatest number of peaks and best selectivity.
- Adjust slope of the chosen gradient by changing initial and final solvent percentages so that selectivity is maintained with a minimum gradient time.
- If the separation of critical peak pairs is not achieved, attempt to include isocratic hold times within the gradient.
- If adequate separation is still not achieved, investigate mix solvents gradient (e.g., ACN/MeOH) holding the solvent strength of the mixture equivalent to the solvent strength chosen for single solvent gradient.

4.1.2 Reference Standards

There are several classes for reference standards as follows:

- Reference Standard, Primary: A substance that has been shown by an extensive set of analytical tests to be authentic material that should be of high purity. This standard can be:

  - 1. Obtained from an officially recognized source.
  - 2. Prepared by independent synthesis.
  - 3. Prepared by further purification of existing material. This standard can be used to certify other material as reference standards. The criteria for a primary reference standard is shown in Table 4.2 below.

Table 4.2 Primary reference standard criteria

Required tests	Acceptance criteria
Enantiomeric purity (Chiral compounds only)	≤0.5 % of the undesirable enantiomer
Residual solvent by GC	≤0.5 % of each individual solvent, unless solvate
Purity by HPLC	≥99.0 % Purity
Counter ion(s)	Report results—consistent with salt stoichiometry
Identity	Must conform to structural identity, via two methods
Water content	≤0.5 %, unless hydrate
Residue on ignition	≤0.5 %
Appearance	Material specific

Table 4.3 Secondary reference standard criteria

Required tests	Acceptance criteria
Residue on ignition (ROI)	≤1.0 %
Enantiomeric purity (Chiral compounds only)	≤2.0 % of the undesirable enantiomer
GC residual solvent	≤1.0 % of each individual solvent, unless solvate
Purity by HPLC	≥ 97.0 % purity
Counter ion(s)	Report result—consistent with salt stoichiometry
Identity	Must conform to structural identity, via two methods
Water content	≤2.0 %, unless hydrate
Appearance	Material specific

Table 4.4 Qualitative reference standard criteria

Required tests	Acceptance criteria
Identity[a]	Conforms to identity by one method
Qualitative examination[b]	Material specific
Appearance	Material specific

[a]A single method either structural (e.g., UV) or physical (e.g., HPLC) is acceptable
[b]An appropriate qualitative test will be defined for the specific molecule (e.g., HPLC purity)

- Reference Standard, Secondary: A substance of established quality and purity, as shown by comparison to a primary reference standard, used as a reference standard for routine laboratory analysis. The criteria is shown in Table 4.3.
- Qualitative Reference Standard. A substance used for qualitative analysis: internal standard, system suitability, marker, racemate. The criteria is shown in Table 4.4.

The storage of the reference standard is an essential component to maintain its validity. It is the responsibility of the recipients to ensure that the reference standards issued are stored and handled under recommended storage condition. Even if the issued reference standard is stored as directed on the label, the extension of the expiration date will not apply to the originally issued material upon re-certification of the reference standard.

4.1.3 Analytical Method Transfer [3]

Analytical method transfer is typically managed under an internal transfer protocol that details the parameters to be evaluated in addition to the predetermined acceptance criteria that will be applied to the results. Transfer studies usually involve two or more laboratories (originating lab and receiving labs) executing the preapproved transfer protocol. A sufficient number of representative test material (e.g., same material) are used by the originating and receiving laboratories. The comparative studies are performed to evaluate accuracy and precision, especially with regard to comparability of interlaboratory variability to within laboratory variability. The industry approach for the transfer of analytical methodology is as follows:

There are three categories for method transfer.

4.1.3.1 Transfer Waiver

- This is done for very simple tests (odor, appearance)
- Method already in the receiving lab (water KF, pH, compendial method, or extensive experience with the method)
- Requires document review
- A Method Transfer Summary Document is written by the receiving laboratory and approved by both the sending and receiving laboratories and quality assurance

4.1.3.2 Method Qualification Transfer

- For qualitative methods (e.g., IR-ID, XRD)
- Generate acceptable results in receiving laboratory
- Requires document review
- The receiving laboratory analysts must be trained on the test method procedures by a qualified laboratory analyst from the sending laboratory, or demonstrate competency
- Testing done in both laboratories to establish the adequacy of implementation of the analytical methods
- Results from the training and testing is examined and a determination if the receiving laboratory is qualified to perform the analytical method
- A Method Transfer Summary Document is written by the receiving laboratory and approved by both the sending and receiving laboratories and quality assurance

4.1.3.3   Validation/Co-validation Transfer

- Concurrent real-time testing in sending and receiving laboratory
- Real-time testing in the receiving laboratory compared to historical results developed in the sending laboratory
- Prepare the Method Transfer Protocol
- Provide the required samples to the sending and receiving laboratories
- The analysts perform the testing as described in the Method Transfer Protocol
- The data is analyzed as required by the protocol and determines if the acceptance criteria have been met
- A Method Transfer Summary Document is written by the receiving laboratory and approved by both the sending and receiving laboratories and Quality Assurance
- Method Transfer Protocol prepared by the sending laboratory in collaboration with the receiving laboratory
- Contains transfer study design, specific lot, acceptance criteria, number of results needed, sample requirements
- Results from sending lab may be available (historical)—must assure are appropriate
- Approvals: sending lab, receiving lab, and QA

Method Transfer Report

- Prepared by the receiving laboratory with input from the sending laboratory
- Contains results, statistical analysis, assessment versus acceptance criteria

Acceptance Criteria

- Precision

  - Upper 95 % confidence interval of the % RSD cannot be exceeded at either site

- Comparison of Means

  - The mean values obtained from the transfer study must be contained within the 95 % confidence interval of means.

4.1.4   Training [1, 3]

4.1.4.1   Purpose

To establish and define an internal training program and to ensure the competency of laboratory personnel. Training and training verification are key factors for successful laboratory operations.

4.1.4.2 Scope

This training procedure is used to ensure that training has taken place with each employee for procedures and methods that the employee performs. The procedure applies to on-the-job training, in-house training, and new-hire training. The training is verified and documented. The training procedure is applicable to new employees, for the introduction of new procedures and methods, for retraining of employees, and for reverification of employee performance.

4.1.4.3 Responsibilities

- Management
  - Ensures implementation of training procedure.
  - Ensures resources are allocated for identified training within budgetary constraints.
  - Responsible for the evaluation, training, and growth of the technical- and quality-related skills of employees by establishing training schedule and rotation for all new employees and by ensuring personnel receive training and demonstrate competence.
  - Ensures training is accomplished.
  - Submits documentation for completed training for entry into training database.
  - Identifies training needs and courses.
  - Implements training and maintains employee training files.
  - Ensures proper supervision of trainees until training is completed.
  - Reviews training received and ensures training files are complete.
  - Submits documentation for training completed.
  - Monitors employee performance to identify the need for retraining or additional continual education.
  - Identifies training needs resulting from new or revised procedures and processes.

4.1.4.4 Quality Management Systems (QMS) Manager

- Trains employees in quality control and quality assurance procedures.

4.1.4.5 Staff

- Completes required training within specified time frame.
- Becomes and stays knowledgeable in procedures and methods performed, *Note*: Employees are responsible for self-training, through reading current literature, technical papers, publishing technical papers.
- Reports all training received and submits documentation for training received.
- Reads and complies with standards, regulations, policies, procedures, and work instructions.

4.1.4.6   Procedure

- Before starting any work-related duties, the employee should be familiar with all work-related documents. These documents include procedures, work instructions, applicable manuals, and regulations.
- Training requirements are outlined and documented on the basis of the position description of duties and responsibilities.
- The level of training is determined by the employee's educational qualifications, experience, complexity of the test method, and knowledge of the test method performed.
- The employee will not perform any procedure, inspection, or method until all applicable training has been completed and competency demonstrated. Employees undergoing training are supervised until training is completed and competency demonstrated.
- Employees may request training related to their job.
- Upon completion of training, submit documentation such as sign-in sheets for entry into training database.
- The effectiveness of training is evaluated by, but not limited to, reviews performed by management and performance evaluations.

4.1.4.7   Training Technique

- The training process for technical procedures such as laboratory analysis consists of the following steps:
  - Trainee reads the laboratory procedures, work instructions, or other applicable documents.
  - Trainee observes demonstration of the procedure by a trainer.
  - Trainee performs the procedure under observation by a trainer.
  - Trainee successfully completes the procedure.
  - Documentation of these tasks is submitted for filing in the employee's training file.
- The training process for nontechnical procedures includes, but is not limited to the following:
  - Reading laboratory and district procedures
  - Instructions
  - Demonstrations
  - Lectures and discussions
  - Self-study
  - Computer-based training
  - Viewing videotapes
  - Documentation of these tasks is submitted for filing in the employee's training file

- Training documentation
- Examples of Training Verification Records include any of the following but is not limited to:

  - Completion of training checklists prepared internally for all procedures that an employee performs
  - Completion of the procedure with submission of written evidence
  - Completion of proficiency surveys, testing with submission of results that are within acceptable criteria limits
  - Completion of written evaluations, signing acknowledgment of reading assigned work instructions
  - Attendance sign-in sheets on in-house training, certificates from courses and computer classes, and committees served on
  - Submission of technical papers and handouts of presentations given, college transcripts for courses taken, licenses and memberships held and special conferences attended
  - Completed paperwork on safety briefing, orientation modules, memorandums on additional appointments or duties

4.1.4.8   Required Training

All analysts and laboratory staff members are to undergo training in a number of procedures, policies, and practices upon entry of employment and during their career. The types of required training are listed below:

- Facility orientation includes:

  - New employees completing required administrative forms as part of initial processing
  - Introduction to co-workers, personnel policies, working conditions, daily routine, issuance of manuals, quality assurance system and any miscellaneous matters.

- New-Hire Training often includes the following:

  - Basic Forensic Drug Law
  - Evidence Development Course
  - Quality Systems and Audit Workshop

- Safety training may include the topics of the following:

  - Hazard communication standard
  - Personal protective equipment
  - Security briefing
  - Radiation protection training
  - Fire extinguisher training
  - Emergency evacuation
  - Safety practices in the laboratory
  - Chemical hygiene

Hazardous Waste Management that includes annual training on handling, storage, and disposal of hazardous materials

Quality assurance including annual training on quality control (QC), quality assurance (QA)

Training on policies, regulations, procedures, methods, and instruments

- Training modules often include the following topics:

  - Basic Analytical Skills
  - Glassware (usage and storage)

    Volumetric
    Graduated

  - Filters
  - Pipetting
  - Syringes
  - Melting Point

    USP
    DSC

  - Thermometers
  - Waste Disposal
  - Calculations

- Laboratory Notebooks Documentation and Storage

  - Issuance
  - Table of Contents
  - Glossary
  - Corrections
  - Cross Reference

- Reporting of Analytical Data

  - Units
  - Rounding
  - Significant figures
  - Sample replicates

- Review and Approval of Data

  - Raw Data

- Electronic Balances

  - Instrument Setup

    Level Balance
    Zero Balance

- Calibration

  Using Weights

- Operation
- Maintenance

• Use of pH Meters

  - Basic Theory
  - Electrodes

    Reference
    Glass
    Combination
    Storage
    Buffers

  - Calibration Procedure
  - Routine Maintenance

• Titration

  - Aqueous
  - Nonaqueous
  - Karl Fischer
  - Oxidation–Reduction
  - Complex–Metric

• Gas Chromatography (GC)

  - Basic Theory
  - Instrument Setup

    IQ, OQ, PQ

  - Instrument Operation
  - Sample Preparation

    Liquid
    Head Space

  - Column Chromatography Phases

    Liquid
    Solid

  - Detection Modes

    FID, NPD, EC
    Tandem (MS, etc.)

  - Troubleshooting, Calibration/Maintenance

- High Performance Liquid Chromatography (HPLC)

  - Basic Theory
  - Instrument Setup

    IQ, OQ, PQ

  - Instrument Operation
  - Sample Preparation
  - Column Chromatography Phases

    Reverse Phase
    Normal Phase
    Ion Exchange
    Size Exclusion
    Gel Permeation
    Vacancy Chromatography
    Chiral
    Affinity

  - Detection Modes

    UV/Visible/Diode Array
    Refractive Index
    Evaporative Light Scattering
    Tandem (MS, etc.)

  - Troubleshooting, Calibration/Maintenance

- Infrared (IR)/Fourier Transform IR Spectroscopy

  - Basic Theory
  - Instrument Setup

    IQ, OQ, PQ

  - Calibration Procedure
  - Instrument Operation
  - Sample Preparation

    KBr Pellet
    NaCl Cells
    Attenuated Total Reflectance
    Nujol Mull
    Diffuse Reflectance Infrared Fourier Transform (DRIFTS)

  - Light Source

    Nernst Glower

  - Detectors

    Quantum
    Thermal

- Ultraviolet/Visible Spectroscopy

  - Basic Theory
  - Instrument Setup

    IQ, OQ, PQ

  - Instrument Operation
  - Sample Preparation

    Quartz Cells

  - Light Source

    Neon Arc
    Tungsten Lamp

  - Monochromators

    Filters
    Prism
    Grating

  - Detectors

    Photomultiplier Tube
    Diode Array
    Calibration and Maintenance

- Thin Layer Chromatography (TLC)

  - Basic Theory
  - Instrumentation
  - Sample Preparation
  - Chromatographic Plates

    Silica Gel (Normal Phase)
    Modified Silica Gel

      KOH (Basic Drugs)
      Octadecylsilane (Reverse Phase)
      Liquid Paraffin (Triglycerides, Fatty Oils)

  - Detection

    Ultraviolet Light
    Fluorescent light
    Quenching
    Location Reagents

      Potassium Permanganate
      Ninhydrin Solution
      Iodine Vapor

Trainer Qualifications—Individuals are considered qualified to perform training after demonstrating expertise in the particular technology. The trainer can also be qualified by an outside training expert. All training activities need to be documented in individual training files.

Grandfathering—Personnel employed prior to the effective date of the training program may be considered qualified or "grandfathered" on techniques based on previous experience or training. This must be documented and in their training file. "Grandfathering" cannot occur once the training program is effective. In the future, all newly hired individuals must go through training irrespective of education and experience.

4.2   Good Laboratory Practices (GLP) [2–4, 6]

4.2.1   Personnel

- Each individual engaged in the conduct of or responsible for the supervision of a laboratory study shall have education, training, and experience, or combination thereof, to enable that individual to perform the assigned functions.
- Each laboratory shall maintain a current summary of training and experience and job description for each individual engaged in or supervising the conduct of a laboratory study.
- There shall be a sufficient number of personnel for the timely and proper conduct of the study according to the testing required

4.2.2   Handling Test and Control Article

- As it is necessary to prevent contamination or mix-ups, there shall be separate areas for:
  - Receipt and storage of the test and control articles.
  - Storage areas for the test and/or control article are separate from laboratory areas and shall be adequate to preserve the identity, strength, purity, and stability of the articles.

4.2.3   Maintenance and Calibration of Equipment

- Equipment shall be adequately inspected, cleaned, and maintained. Equipment used for the generation, measurement, or assessment of data shall be adequately tested, calibrated, and/or standardized.

- A written standard operating procedures shall be written in sufficient detail the methods, materials, and schedules to be used in the routine inspection, cleaning, maintenance, testing, calibration, and/or standardization of equipment, and shall specify, when appropriate, remedial action to be taken in the event of failure or malfunction of equipment. The written standard operating procedures shall designate the person responsible for the performance of each operation.
- Written records shall be maintained of all inspection, maintenance, testing, calibrating, and/or standardizing operations. These records, containing the date of the operation, shall describe whether the maintenance operations were routine and followed the written standard operating procedures. Written records shall be kept of non-routine repairs performed on equipment as a result of failure and malfunction. Such records shall document the nature of the defect, how and when the defect was discovered, and any remedial action taken in response to the defect.

4.2.4 Reporting of Results [5]

- Final report shall be prepared for each laboratory study and shall include, but not necessarily be limited to, the following:
  - Name and address of the facility performing the testing and the dates on which the testing was initiated and completed
  - Statistical methods employed for analyzing the data.
  - All test and control articles identified by name, chemical abstract number, strength, purity, and composition or other appropriate characteristics.
  - A description of the transformations, calculations, or operations performed on the data, a summary and analysis of the data, and a statement of the conclusions drawn from the data analysis.

4.2.5 Standard Operating Procedure

A testing facility shall have standard operating procedures in writing for laboratory methods that are adequate to insure the quality and integrity of the data generated in the course of a study. All deviations from standard operating procedures shall be authorized by the quality assurance and shall be documented in the raw data. Significant changes in established standard operating procedures shall be properly authorized in writing by management. Standard operating procedures shall be established for, but not limited to, the following:

An example of SOPs which are commonly written are given below:

4.2.5.1 Laboratory Operations

- Automated Instrument Implementation and Use
- Rounding Rules
- Decimal Place Reporting for Analytical Data
- Replicate and Composite Size Determination for sample testing
- Reporting Impurities, Including Degradation Products
- Analytical Data Review Process
- Equipment Calibration and Maintenance
- Receipt, identification, storage, handling, and method of sampling of the test and control articles.
- Each laboratory area shall have immediate availability to laboratory manuals and standard operating procedures

4.2.5.2 Department Policies

- Notebook/Data Handling/Creation and Use of Work Sheets
- Analysis Request/Sample Handling/Reports of analysis
- Data handling, storage, and retrieval.
- Analytical Method Development
- Analytical Method Document
- Method Validation Reports
- Characteristics of Method Validation
- System Suitability for Chromatographic Methods
- Retention Sample Policy
- Records Retention
- Reference Standard Certification

4.2.5.3 Management Control

- Documents for submission to official agencies
- A historical file of standard operating procedures, and all revisions, including the dates of the revisions, shall be maintained
- Method Transfer Process
- Personnel Training and Certification program

4.2.6 Reagents and Solution labeling

- All reagents and solutions in the laboratory areas shall be labeled to indicate identity, strength, concentration, storage requirements, and expiration date. Outdated reagents and solutions shall not be used.

4.2.7 Conduct of Laboratory Study

- All data generated during a laboratory study, except those that are generated by automated data collection systems, shall be recorded directly, promptly, and legibly in ink.
- All data entries shall be dated on the date of entry and signed or initialed by the person entering the data.
- Any change in entries shall be made so as not to obscure the original entry, changes shall indicate the reason for such change, and shall be dated and signed or identified at the time of the change.
- In automated data collection systems, the individual responsible for direct data input shall be identified at the time of data input.
- Any change in automated data entries shall be made so as not to obscure the original entry, shall indicate the reason for change, shall be dated, and the responsible individual shall be identified.

4.2.8 Retention of Records

- All documentation records, raw data to a laboratory study shall be retained in the archive(s) for whichever of the following periods is shortest.

  - Period of at least 5 years following the date on which the results of the laboratory study are submitted in support of the intended use of the study.
  - Copies of protocols, and records of quality assurance inspections, shall be maintained by the quality assurance unit for accessibility of records for the period of 5 years.
  - Records and reports of the maintenance and calibration and inspection of the equipment shall be retained for the lifetime of the equipment.
  - Records required may be retained either as original records or as true copies such as photocopies, microfilm, microfiche, or other accurate reproductions of the original records.

References

1. 21CFR, 210, 211 (2006) Pharmaceutical cGMP regulations
2. 21CFT, 58 (2006) Pharmaceutical cGLP regulations
3. Catalano T (2013) Essential elements for a GMP analytical chemistry department. Springer, New York
4. National Research Council (2009) Strengthening forensic science in the United States, a path forward. National Academies Press, Washington, DC
5. Inman K, Rudin N (2001) Principles and practice of criminalistics. CRC, Boca Raton
6. Shelton DE (2011) Forensic science in court challenges in the twenty-first century. Rowman & Littlefield, Lanham
7. Hadley K, Fereday MJ (2008) Ensuring competent performance in forensic science. CRC, Boca Raton

Chapter 5
International Committee on Harmonization (ICH)

The ICH guidelines consist of sections Q1–Q11. These guidelines are complementary to the GLPs and GMPs; however, they describe in more detail the activities to be performed and the criteria to be achieved. Although these are guidelines, they should be followed when generating data, and writing justifications, protocols and reports. If not followed, a very strong justification must be submitted to the agency along with comparative data or logic [1–3].

5.1 Validation of Analytical Procedures [1, 4]

The objective of validation of an analytical procedure is to demonstrate that it is suitable for its intended purpose. The characteristics applicable to identification, control of impurities, and assay procedures are described in detail.

- Identification—Tests intended to ensure the identity of an analyte in a sample. This is normally achieved by comparison of a property of the sample (e.g., spectrum, chromatographic behavior, chemical reactivity,) to that of a reference standard. It also required that the test is specific for the analyte and in many cases requires two independent tests to ensure the analyte identification.
- Testing for impurities—These tests can be either a quantitative test or a limit test for the impurity in a sample. Either test is intended to accurately reflect the purity characteristics of the sample. Different validation characteristics are required for a quantitative test than for a limit test.
- Assay procedures—Intended to measure the analyte present in a given sample. The assay represents a quantitative measurement of the major component(s) in the sample. Similar validation characteristics also apply when assaying for the other selected component(s).

© Thomas Catalano 2014
T. Catalano, *Good Laboratory Practices for Forensic Chemistry*,
SpringerBriefs in Pharmaceutical Science & Drug Development,
DOI 10.1007/978-3-319-09725-1_5

- The objective of the analytical procedure should be clearly understood since this
  will govern the validation characteristics which need to be evaluated. Typical
  validation characteristics which should be considered are listed below:
  - Accuracy
  - Precision
  - Repeatability
  - Intermediate precision
  - Specificity
  - Detection limit
  - Quantitation limit
  - Linearity
  - Range
  - Robustness (should be performed during development)

5.1.1   Specificity

- An investigation of specificity should be conducted during the validation of identification tests, the determination of impurities, and the assay. The procedures used to
  demonstrate specificity will depend on the intended use of the analytical procedure.
  It is not always possible to demonstrate that an analytical procedure is specific for a
  particular analyte. In this case, a combination of two or more analytical procedures
  is recommended to achieve the necessary level of discrimination.
- Method specificity can be demonstrated by the force degradation of the analyte
  under various conditions (e.g., temperature, humidity, acid, base, oxidation, and
  light) and by the introduction of know impurities or closely related substances
  which may be present with the analyte.
- The use of peak purity tests to show that the analyte is composed of only component
  is recommended (e.g., diode array, mass spectroscopy).

5.1.2   Linearity

- The linearity of the analytical method should be investigated. This can be demonstrated by serial dilution of the reference material over the expected concentration
  range. Linearity should be evaluated by visual inspection of a plot of the response
  signal versus the concentration of the analyte. If the data appears to be linear, the
  data should be evaluated by appropriate statistical analysis, such as linear least
  squares regression analysis. This analysis can help evaluate linearity by calculating
  the correlation coefficient, y-intercept, and the slope of the line. In addition the calculation of the deviation of the actual data points from regression line (residuals) is
  very helpful in evaluation whether the acceptable linearity has been accomplished
  for the methods' intended use. The establishment of linearity should consist of the
  evaluation of a minimum of five concentrations over a specified range.

5.1.3 Accuracy

- Accuracy is determined by the application of the analytical method to known quantities of the analyte (e.g., reference material).
- Application of the analytical method to known mixtures of the analyte with components of which may be found with the analyte.
- Accuracy may be inferred once linearity, specificity, and precision have been established.
- Quantitation of impurities can be determined by spiking known amounts of impurities with the analyte or if impurities are not available by evaluating the analyte at the expected impurity concentrations.
- Accuracy should be assessed using a minimum of nine determinations over a minimum of three concentration levels covering the specified range (e.g., three concentrations/three replicates each).
- Accuracy should be reported as percent recovery of known added amount of analyte in the sample.

5.1.4 Precision

- The investigation of precision is required when the validation of assay and the quantitative determination of impurities are being considered.
- Repeatability should be performed using a minimum of six replicate determinations at the 100 % target level.
- Intermediate precision should be determined based on the intended use of the method. The effects of random events on the precision of the method should be considered. Variations such as different runs, different analyst, and different equipment should be considered.
- Reproducibility is the comparison of precision data from an interlaboratory trial. Reproducibility can be substituted for intermediate precision.
- The recommended data to report for each type of precision is the standard deviation, relative standard deviation, and their 95 % confidence intervals.

5.1.5 Quantitation Limit

- The Quantitation Limit (LOQ) is the lowest concentration level which can be quantitated with acceptable accuracy and precision.
- The determination of the quantitation limit for the analytical method can be done in several ways depending on whether the method is an instrumental or non-instrumental.
- The LOQ can be determined by measuring the ratio of the signal of the analyte at a low concentration and the noise of the baseline of a blank sample. A typical signal to noise ratio is 10:1.

- The LOQ can also be determined utilizing standard deviation σ of the baseline noise. To normalize this signal to noise ratio with detectors of different responses, the standard deviation is divided by the slope of the calibration curve.

$LOQ = 10\sigma/S$

- The LOQ is subsequently validated by the replicate analysis of the sample with known concentration near the quantitation limit and obtaining acceptable accuracy and precision.

5.1.6 Detection Limit

- The Detection Limit (LOD) can be determined using the same approach used for the determination of the LOQ. Typical signal to noise ratio is 3:1.

$LOD = 3.3\sigma/S$

- Where an estimated value for the detection limit is obtained by calculation or extrapolation, this estimate may be validated by the analysis of a number of replicate samples near the detection limit.

5.1.7 Robustness

- The evaluation of robustness should be considered during the development. It should show the reliability of an analysis with respect to deliberate variations in method parameters.
- If the results of the robustness study indicate that unacceptable variation occurs, a precautionary statement should be included in the analytical method.
- An outcome from the robustness study is the determination of a series of parameters that must be controlled to ensure the validity of the analytical method is maintained each time it is implemented.
- Examples of typical variations utilized in the robustness study are:
- Stability of analytical solutions

  - Extraction time
  - Variations of pH in a mobile phase
  - Variations in mobile phase composition
  - Different columns (different lots and/or suppliers)
  - Temperature
  - Flow rate

5.2   Quality Risk Management [1]

Quality risk management is a systematic process for the assessment, control, communication, and review of risks to the quality of the data produced through the life cycle of the study.

- Quality risk management is a process designed to coordinate and improve science-based decision making with respect to risk, such as

  - Defining the problem or risk question
  - Obtaining background information or data on the consequence of the potential risk
  - Identifying critical resources and a team leader
  - Specifying a timeline, deliverables, and the appropriate level of decision making for the identified risks.

- Some of the simple techniques that are commonly used to structure risk management by organizing data and facilitating decision making are:

  - Flowcharts
  - Check Sheets
  - Cause and Effect Diagrams (also called the fish bone diagram)

- Risk assessment consists of the identification of consequences and the analysis and evaluation of risks associated with exposure to those consequences.
- Quality risk assessments begin with a well-defined problem description or risk question. When the risk in question is well defined, the risk management tool and the types of information that will address the risk question can be identifiable. To clearly define the risk(s), three fundamental questions are often asked:

  - What might go wrong?
  - What is the likelihood (probability) it will go wrong?
  - What are the consequences (severity)?

- Risk evaluations compare the determined risk against a given risk criteria. Risk evaluations are based on the strength of evidence for the three fundamental questions.
- For an effective risk assessment, the robustness of the data set is important because it determines the quality of the data. Revealing assumptions and reasonable sources of uncertainty will enhance confidence in the data and identify its limitations. Uncertainty is due to combination of incomplete knowledge about data production and its variability. Typical sources of uncertainty include gaps in knowledge, gaps understanding, sources of method variability, and the probability of detecting the problems.
- The output of a risk assessment is either a quantitative estimate of risk or a qualitative description of a range of risk. When risk is expressed quantitatively, a numerical probability is attached. Alternatively, risk can be expressed using qualitative descriptors, such as high, medium, or low, which should be defined in

as much detail as possible. Sometimes a *risk score* is used to further define descriptors in risk ranking. In quantitative risk assessments, a risk estimate provides the likelihood of a specific consequence, for a risk circumstances. Thus, quantitative risk estimation is useful for one particular consequence at a time. Alternatively, some risk management tools use a relative risk measure to combine multiple levels of severity and probability into an overall estimate of relative risk. The steps in the scoring process can employ quantitative risk estimation.

- Risk control includes decision making to reduce and/or accept risks. The purpose of risk control is to reduce the risk to an acceptable level. The amount of effort used for risk control should be proportional to the significance of the risk. Risk control might focus on the following questions:

    - Is the risk above an acceptable level?
    - What can be done to reduce or eliminate risks?
    - What is the appropriate balance among benefits, risks, and resources?
    - Are new risks introduced as a result of the identified risks being controlled?
    - Risk communication is the sharing of information about risk and risk management between the decision makers and others. Parties can communicate at any stage of the risk management process. The output/result of the quality risk management process should be appropriately communicated and documented. Communications might include those among interested parties (e.g., laboratory management, prosecutors, defense, and courts). The information communicated might relate to the existence, probability, severity, acceptability, control, and detectability, or other aspects of risks to quality.

5.3 Quality Assurance System [1]

A quality assurance system support implementation of an effective quality system to enhance the quality of forensic data generated in the interest of public safety. Implementation of quality assurance throughout the study life cycle should facilitate continual improvement and strengthen the link between prosecutors, defense attorneys, and the courts, by establishing the reliability of the data and its interpretation.

- Implementation of the quality assurance model should result in achievement of major objectives that complement or enhance demanded requirements.

    - Establish and maintain a state of control by developing an effective monitoring and control systems for data quality, thereby providing assurance of continued suitability of the data. Quality risk management can be useful in identifying the monitoring and control systems.
    - The design, organization, and documentation of the quality system should be well structured and clear to facilitate common understanding and consistent application.

- Management responsibilities should be identified within the quality assurance system.
- The quality system should include the following elements, data quality monitoring, corrective and preventive action, change control management, and management review.

• A quality policy should be developed and contain the following elements:

- Senior management should establish a quality policy that describes the overall intentions and direction of the laboratory related to quality.
- The quality policy should include an expectation to comply with applicable federal, state, and local requirements and should facilitate continual improvement of the quality assurance system.
- The quality policy should be communicated to and understood by personnel at all levels in the laboratory.
- The quality policy should be reviewed periodically for updating to current standards.
- Senior management should be responsible for the quality system governance through management review to ensure its continuing suitability and effectiveness.

References

1. International Committee on Harmonization (ICH), Q2(R1), (2005), Q7A(2001), Q9(2006), Q10(2009), U.S. Food and Drug Administration: Silver Spring
2. Catalano T (2013) Essential elements for a GMP analytical chemistry department. Springer, New York
3. Hadley K, Fereday MJ (2008) Ensuring competent performance in forensic science. CRC, Boca Raton
4. Inman K, Rudin N (2001) Principles and practice of criminalistics. CRC, Boca Raton

Chapter 6
International Organization
for Standards (ISO)

ISO 17025:2005 is the first internationally accepted standard for laboratory quality systems. It contains two main sections, namely Management Requirements and Technical Requirements. To have an efficient and effective laboratory, you will need to develop a management team as well as a technical team. ISO 17025 precisely addresses the roles and responsibilities of management as well as the technical requirements [1].

6.1 Management Requirements (Clause 4)

- The first item to be addressed is the development of an organization that has a legal identity, type of organization, and the scope of activities described in detail. Clause 4.1.
- The management system should develop a quality manual based on the objectives of the laboratory. The systems within the manual should adhere to the content in clause 4.2.
- The implementation of the management system requires the documentation of the system and its procedures, such as document control, writing and issuance of the document, and changes to the documents. This is detailed in clause 4.3.
- In the course of normal business, there is always the probability of obtaining nonconforming data within the laboratory testing. Thus it is important to have policies and procedures to investigate nonconforming data, calibrations, laboratory errors, etc. This is described in clause 4.9.
- After the laboratory investigation is complete, any assignable cause, if found, should be identified, along with a root cause and a corrective action. Details are found in clause 4.11.

© Thomas Catalano 2014
T. Catalano, *Good Laboratory Practices for Forensic Chemistry*,
SpringerBriefs in Pharmaceutical Science & Drug Development,
DOI 10.1007/978-3-319-09725-1_6

- After a nonconformity is identified and reported, a system to describe preventive action takes place to eliminate the reoccurrence of the nonconformity in the future.
- In the operation of a laboratory, there are many records generated over the course of time. A system of policies and procedures to control the documents for easy and timely retrieval must be available. All documents which have several revision must be controlled using a document revision control system. This is described in clause 4.13.
- In order to ensure that the laboratory is in compliance with the Quality Management System, periodic internal audits of the laboratory against the Quality Manual must be performed and the results communicated with the laboratory personnel and Sr. management. All negative audit finds must be corrected as described in the Quality Manual. See clause 4.14.
- Periodic training (usually annually) is required to ensure both management and staff are keep up to date on all changes, updates, new requirements, etc. These training sessions should be open and interactive; minutes should be taken and placed into document control. Clause 4.15.

6.2   Technical Requirements (Clause 5)

- Management requirements are extremely important in the development and operation of a laboratory organization. However, technical requirements are equally important if not more important since that are required for the implementation of the laboratory technology. Each laboratory operation or job description should include the technical requirements needed to fulfill the activities required. A detailed list of these activities should be available, since they would contribute to measurement of uncertainty that would evaluated during activities such as methods development, training, qualification of personnel, and the calibration of equipment.
- The first technical component of the any laboratory is the quality of the technical staff. The technical personnel must be suitably qualified, through education and training, for the job duties they will be performing. In addition the laboratory must have an approved process for the ongoing training for all job descriptions and responsibilities required of the technical personnel. See clause 5.2.
- The laboratory should have suitable environmental conditions for the staff to carry out the laboratory testing activities. The facility should have acceptable housekeeping procedures that are followed on a daily basis with inspections performed on a routine basis. The facility should have controlled access with identified areas to provide separation between incompatible activities. Necessities such as adequate utilities, lighting, temperature, humidity, and control should be provided so that the laboratory operations can be carried out accurately. See clause 5.3.

- All testing requires the use of suitable methods. These can be standard compendial methods, methods from the scientific literature, or methods developed in-house. Method developed in-house or literature methods must be suitability validated. All methods should incorporate criteria within the method to control the quality of the data generated. See clause 5.4.
- All equipment should be maintained and calibrated on a regular basis according to an approved plan. Each piece of equipment should have an identified person responsible for the maintenance and calibration. See clause 5.5.
- The sampling strategy for obtaining the test sample is of utmost importance to generate data that is representative of the original bulk material. A suitable sampling procedure should be in place and should be based on a statistical approach. How the sampling was preformed should document for each test sample utilized. See clause 5.7.
- A sample receiving and distribution process should be in place, so that a sample chain of custody and integrity is maintained and traceable. The process should include the handling, storage, retention, and destruction throughout the life of sample in the laboratory. The laboratory should have suitable facilities containing multiple conditions (e.g., room temperature with controlled humidity, refrigeration, freezers) to avoid sample degradation. The sample receiving person should document an abnormalities with sample upon accepting the sample into the laboratory. See. Clause 5.8.
- The assurance of the validity and reliability of the data generated is of the utmost importance in a forensic environment. This can be done by having quality criteria within the analytical procedure such as a required precision of the injections, a required standard deviation for all reference standard data, a difference limit for replicated results from the same sample. See clause 5.9.
- Reporting of the data in an Analysis Report can have different formats depending on the type of analysis performed. However, there is essential information that should be reported regardless of the report format. The essential information required are listed below: See clause 5.10.

  - Report title
  - Name and address of laboratory
  - Project number
  - Unique identifier for the sample
  - Report number and version
  - Sample and Standards Lot numbers
  - Sample description
  - Types of analyses requested
  - Sample submitter's name
  - Analytical method number and revision
  - Reporting units (%, mg/mL, ppm, etc.)
  - Test results, individual, mean, standard deviation

- Reporting significant figures
- Analyst's signature and date
- Review analyst's signature and date
- Approving supervisor's signature and date

Reference

1. International Organization for Standards (ISO) (2005) Clause 4 management requirements, clause 5, technical requirements. ISO, Geneva

Chapter 7
Statistical Considerations

7.1 Normal Distribution [1]

When analytical data is collected and plotted against a variable, a distribution of the data is observed. If the experiments are repeated and the data plotted, the distribution approaches a form which can be described by a mathematical equation. Most analytical data will approach a distribution function which is described as a "Normal Distribution." The normal distribution arises from the summation of many small random errors and can be expressed mathematically as $\int_{-\infty}^{\infty} \frac{1}{\sigma\sqrt{2\pi}}\left[e^{-\frac{(x-\mu)^2}{2\sigma^2}}\right]dx = 1$.

It can be seen from the expression that the distribution is completely determined by the parameters μ the mean and the variance σ^2. The distribution is symmetrical about the mean, where the highest probability of measurements will occur, the probability of measurements occurring drops off sharply with 68.26 % occurring at $\mu \pm \sigma$, 95.44 % at $\mu \pm 2\sigma$, and 99.74 % at $\mu \pm 3\sigma$.

7.2 Significance [1, 3, 4]

- The student t distribution is routinely used for determining if results have significant bias or for comparing observations with limits. The distribution is symmetrical and resembles a normal distribution and is described by the parameter "Degrees of Freedom (v)." As the number of degrees of freedom increases, the distribution approaches the normal distribution. The t distribution is also used for determining confidence intervals. The degrees of freedom refers to the number of independent pieces of data that has been used to measure a particular parameter. In general the degrees of freedom is based on the sample number (n) minus the

© Thomas Catalano 2014
T. Catalano, *Good Laboratory Practices for Forensic Chemistry*,
SpringerBriefs in Pharmaceutical Science & Drug Development,
DOI 10.1007/978-3-319-09725-1_7

number of parameters estimated from the data, for example, if the calculation of the standard deviation requires the determination of the mean (x), the degrees of freedom utilized for the calculation of the standard deviation is $n - 1$.

- The F distribution describes the ratio of the variances. This is important when comparing the difference of variance of two data sets, it is also used in the analysis of variance (ANOVA). It is also used for comparing the precision of alternate methods of analysis. The F distribution is a ratio of two variances and is characterized by the degrees of freedom for each. The asymmetry of the distribution increases as the respective degrees of freedom decreases.

- If two sets of data are determined to have different means X_a and X_b, it is possible that both data sets come from the same population and that the difference observed was due to just random variation in the data generated. However, it is also possible that data generated came from different populations and their means are truly different and not due to random variations. Significant testing can provide an approach which will allow for deciding which is the more likely alternative. Some basic statistic parameters are needed to be determined in order for significant testing to be performed.

- Arithmetic Mean

  The mean X is the summation of all observations divided by the number of observations

$$\bar{x} = \frac{\sum_{i=1}^{n} X_i}{n} \tag{7.1}$$

- Median

  The median is the central member of a series of observations arranged in ascending order. The median will have equal numbers of observations smaller and larger than its value. The median is generally a more robust value since it is less affected by extreme values, such as outliers. For example, the series of observation 1, 5, 7, 8, 11, 12, 15, 20, 25 has a median value of 11.

- Standard Deviation

  The variance within a data set is the mean squared deviation of the values from the data. The standard deviation is the square root of the variance. The variance and standard deviation are estimates of how the values in the data set differ from each other; the larger the variance or standard deviation, the larger the spread of data points within the data set.

$$\text{Variance}\left(S^2\right) = \frac{\sum_{i=1}^{n}\left(x_i - \bar{x}\right)^2}{n - 1} \tag{7.2}$$

$$\text{Standard Deviation}\left(S\right) = \sqrt{\frac{\sum_{i=1}^{n}\left(x_i - \bar{x}\right)^2}{n - 1}} \tag{7.3}$$

It is important to know that when data is being considered for pooling, the standard deviation of each estimate must be squared before summation.

- Standard Error of the Mean
  The standard error of the mean represents the variation of the mean. It represents the uncertainty which occurs from the random variation within an experiment. It is more accurate than the standard deviation because it estimates the variation of averages.

$$\text{Standard Error of the Mean } S\left(\overline{x}\right)=\frac{S}{\sqrt{n}} \qquad (7.4)$$

- Relative Standard Deviation (Coefficient of Variation)
  The relative standard deviation (RSD) or coefficient of variation (CV) is a comparison of the mean with spread in the data. The RSD is one of the most commonly used statistical parameters in analytical chemistry; it is usually expressed a percentage.

$$\%\text{RSD} = \%\text{CV} = \frac{S}{\overline{x}}\times100 \qquad (7.5)$$

- Hypotheses
  The null and the alternate hypotheses are the questions which must be asked when evaluating the significance of the data. The null hypothesis, which denoted as H_0, is interpreted as there is no difference between the data being compared and that the data is from the same population and any difference is from random variation. The null hypothesis is expressed as follows:

$$H_0 : u_A = u_B \qquad (7.6)$$

where u = the population mean
The alternate hypothesis, denoted by H_1, is interpreted as the data being different is from different data populations and the difference is not just due to random variation. The alternate hypothesis is expressed as follows:

$$H_1 : u_A \neq u_B \qquad (7.7)$$

- Significant testing
  When performing significant testing, the following should be considered.

  1. State the null and alternate hypotheses.
  2. Select the appropriate test statistic. In analytical chemistry the most common test statistics utilized are the t-test and the F-test. The t-test is used for the comparison of means and the F-test for the comparison of variances. In each test the calculated test statistic is compared to a critical value obtained from their distribution table at a particular level of probability for a determined level of degrees of freedom.

3. Choose whether the test statistic will be one-tailed or two-tailed. If the alternate hypothesis is H_1: $u_A \neq u_B$, then we are considering a two-tailed statistic since we are concerned if there is a significant difference for the mean of the data set in either directions. If we only want to consider whether the mean of one data set is significantly greater than the mean of the other data set, the alternate hypothesis is H_1: $u_B > u_A$ or if the mean is significantly less than the mean of the other data set H_1: $u_B < u_A$, then we would consider a one-tailed statistic.

4. Choosing the level of significance is determined by what level of probability (α) is acceptable for difference between the data to be considered due to non-random error and be considered significantly different. In analytical chemistry the most common level of probability chosen is $\alpha = 0.05$ level. This would suggest that the probability of values occurring at <0.05 of the distribution would be considered to reject the null hypothesis and be considered a significant difference. If allowing 5 % of the values to reject the null hypothesis is too large, then a lower value such as 0.01 should be used.

5. The t-statistic can be used for the following comparisons:

    (a) The comparison of the mean of a data set to a set value (one sample t-test)
    (b) The comparison of means from two independent data sets (two sample t-test)

    One sample t-test calculation

$$t = \frac{|\bar{x} - u_0|}{S/\sqrt{n}} \tag{7.8}$$

$H_1 : u \neq u_0$, two-tailed

$$t = \frac{\bar{x} - u_0}{S/\sqrt{n}} \tag{7.9}$$

$H_1 : u > u_0$, one-tailed

$$t = \frac{u_0 - \bar{x}}{S/\sqrt{n}} \tag{7.10}$$

$H_1 : u < u_0$, one-tailed
Two sample t-test calculation

$$t = \frac{|x_1 - x_2|}{S_{\text{diff}}}$$

$H_1 : u_1 \neq u_2$, two-tailed

$$t = \frac{x_1 - x_2}{S_{\text{diff}}}$$

$H_1 : u_1 > u_2$, one-tailed

$$t = \frac{x_2 - x_1}{S_{\text{diff}}}$$

$H_1 : u_1 < u_2$, one-tailed

However, the two sample t-statistic differs from the one sample t-statistic in that there are two data sets and therefore two standard deviations. Since there are two standard deviations one from each data set, the calculation requires the pooling of the standard deviations. If each data set has an equal number of observations, the pooled standard deviations (S_{diff}) can be simplified to

$$S_{\text{diff}} = \sqrt{\frac{S_1^2 + S_2^2}{n}} \tag{7.11}$$

If the data sets have unequal amounts of observations, then the calculation of the two sample t-statistic becomes more complicated and statistician should be consulted.

6. Once the t-statistic is calculated it is compared to the critical value obtained from the t-distribution table at the chosen probability level, usually 0.05, with the degrees of freedom $n-1$ for the one sample t-test and the degrees of freedom of $n_1 + n_2 - 2$ for the two sample t-test.

7. Another important statistic is the F-test statistic. The F-test is used for comparing variances s_a^2 and s_b^2 from two independent sets of data. Since variances are squared standard deviations, it can also be used for comparing the precision of analytical methods to see if one is significantly better than the other. The F-test statistic is calculated as a ratio of the variances s_a^2 and s_b^2 with larger value in the numerator and the smaller in the denominator. Since we are comparing variances from two independent data sets, we need to utilize the degrees of freedom (v_a, v_b) for each of the variances. The F distribution table has the degrees of freedom listed on the top row for the numerator and the degrees of freedom for the denominator down the left side of the table. The calculation of the F-test statistic is shown below. The F = test statistic can be used to compare an observed variance to an expected or required variance. In this circumstance the F-test calculation utilized is $F = \dfrac{S_b^2}{S_a^2}$ with the degrees of freedom $v_a = n - 1$ and the degrees of freedom $v_b = \infty$.

F-test calculation

$$F = \frac{\sigma^2_{max}}{\sigma^2_{min}}$$

$$H_1 : \sigma^2_a \neq \sigma^2_b \quad v_{max}, v_{min}$$

(7.12)

$$F = \frac{S^2_a}{S^2_b}$$

$$H_1 : \sigma^2_a > \sigma^2_b \quad v_a, v_b$$

(7.13)

$$F = \frac{S^2_b}{S^2_a}$$

$$H_1 : \sigma^2_a < \sigma^2_b \quad v_b, v_a$$

(7.14)

7.3 Confidence Intervals [1]

The confidence interval is a range of values which would include the mean $\bar{x}$ with a given level of confidence. The 95 % confidence interval about the observed mean $\bar{x}$ is described by the following equation:

$$u - 1.96 \left(\frac{\sigma}{\sqrt{n}} \right) < \bar{x} < u + 1.96 \left(\frac{\sigma}{\sqrt{n}} \right)$$

(7.15)

The value 1.96 comes from the two-tailed t value at $\alpha = 0.05$ and $v = \infty$. However, analytical data usually consist of a relatively small number of data points which is used to calculate the mean $\bar{x}$. Therefore the population standard deviation σ and mean u is not known and must be replaced with the standard deviation S, and the mean $\bar{x}$. The 1.96 is replaced with two-tailed t-statistic value with $n-1$ degrees of freedom. The confidence interval for the mean equation is then revised to

$$\bar{x} \pm t \left(S / \sqrt{n} \right)$$

(7.16)

7.4   Outlier Data [1]

An outlier is an individual data point that is not consistent with rest of the data set. The outlier is usually observed as being distant from the remainder of the data set. These values have a large effect on calculated mean values and standard deviation values. Random variation can occasionally generate extreme values which are part of the valid data set and should be included in the data calculations. However, extreme data can also be produced by human error, analytical procedure error, and instrument failure. These types of outliers should not be included in the final results, so as to reduce their impact on the conclusions.

Outlier testing can be performed to identify the outliers and determine if they are due to random variation or due to some bias. Visual inspection of the data will usually detect suspected outliers. The identification of an outlier by using outlier tests allows the analyst to direct their attention to problems and provide objective criteria for performing inspections or corrective actions. Outliers should not be removed from the data set solely on the bases of a statistical test; the decision should be based on the statistic and on an investigation procedure such as a corrective action and preventive action (CAPA). It is important to realize that outliers are only relative to what is expected in the data set. It is important to consider that the outlier may be a relevant part of the data population. For example, a granular sample where some particles have a greater content of the analyte than most of the other particles. In this case the outlier value is a relevant value in the data set and must be included. Statistical testing for outlier is usually tested at both the 95 and 99 % confidence level. Outliers significant at the 99 % confidence level are usually rejected from the data set; however, rejection of a large portion of the data is not permissible. Outliers significant at the 95 % confidence level are generally not rejected unless supported by other technical reasons.

Most analytical data sets are relatively small <50 observations; therefore, the Dixon Q Test is commonly utilized as an outlier test. Applying the Dixon Q Test begins with ranking all of the data in ascending order $x_1, x_2, x_3, \ldots, x_n$. Calculate the Q statistic for both the high and low outliers as described in Table 7.1:

When applying the Dixon Q statistic, the recommended ranges of the data set sizes should be followed, however going slightly beyond the recommended data set size is not usually serious. As the data set size increases, there is a probability of two outlier values of masking each other. Following the recommend applications for each test will help avoid the masking of outliers and increase the probability of finding these aberrant values. Using several Q-test statistics on the same data set is not recommended.

Table 7.1 Dixon Q test

Test statistic	Application	Q table column (r_i) $i = 10, 11, 12, 20, 21, 22$
$\dfrac{x_2 - x_1}{x_n - x_1}$ (low value)	Test for a single outlier value in a data set of $n = 3\text{--}7$	r_{10}
$\dfrac{x_n - x_{n-1}}{x_n - x_1}$ (high value)		
$\dfrac{x_2 - x_1}{x_{n-1} - x_1}$ (low value)	Test for a single outlier value in a data set, unaffected by a single outlier value at the other end of the data set. Used for a data set of $n = 8\text{--}10$	r_{11}
$\dfrac{x_n - x_{n-1}}{x_n - x_2}$ (high value)		
$\dfrac{x_2 - x_1}{x_{n-2} - x_1}$ (low value)	Test for a single outlier value in a data set, unaffected by up to two outlier values at the other end of the data set. Used for a data set of $n = 5\text{--}10$	r_{12}
$\dfrac{x_n - x_{n-1}}{x_n - x_3}$ (high value)		
$\dfrac{x_3 - x_1}{x_n - x_1}$ (low value)	Test for a single outlier value in a data set, unaffected by one adjacent outlier value. Used for a data set of $n = 5\text{--}10$	r_{20}
$\dfrac{x_n - x_{n-2}}{x_n - x_1}$ (high value)		
$\dfrac{x_3 - x_1}{x_{n-1} - x_1}$ (low value)	Test for a single outlier value in a data set, unaffected by one adjacent outlier value or an outlier value at the other end of the data set. Used for a data set of $n = 11\text{--}13$	r_{21}
$\dfrac{x_n - x_{n-2}}{x_n - x_2}$ (high value)		
$\dfrac{x_3 - x_1}{x_{n-2} - x_1}$ (low value)	Test for a single outlier value in a data set, unaffected by one adjacent outlier value or up to two outlier values at the other end of the data set. Used for a data set of $n = 14\text{--}30$	r_{22}
$\dfrac{x_n - x_{n-2}}{x_n - x_3}$ (high value)		

Reprinted from [6] permission Springer Science + Business

7.5   Linear Regression [2]

Linear regression analysis is utilized to determine the relationship between two variables in a data set. The most common comparison is between the concentration of an analyte and its response from an analytical technique. For example, plotting the response from an analyte (dependent variable) on the y-axis and its concentration (independent variable) on the x-axis the plot will display the relationship between each variable. The purpose of regression analysis is to define the relationship in terms of a mathematical equation. If the relationship is believed to be linear, the equation can be written as

$$y = a + bx \qquad (7.17)$$

where b is the slope of the line and a is the intercept on the y-axis. The method of least-squares linear regression is use to determine the values of a and b for the best fitted line to the data. The best fitted line from the least-squares linear regression is determined by minimizing the sum of the squared differences between the observed values and the fitted values of y. The difference between the observed value and the fitted value ($\hat{y}$) is known as the residual. Before carrying out the least-squares linear regression calculation, a visual examination of the data should be performed. A scatter plot of the data should be prepared and examined for the appearance of an outlier or a disproportionate spread of the data. Either of these conditions could have a significant effect on the position of the regression line, and affect the values of the slope and intercept of the regression line. The calculations of the slope b, and the intercept a, are as follows:

$$b = \frac{\sum\limits_{i=1}^{n}\left[(x_i - \bar{x})(y_i - \bar{y})\right]}{\sum\limits_{i=1}^{n}(x_i - \bar{x})^2} \qquad (7.18)$$

$$a = \bar{y} - b\bar{x} \qquad (7.19)$$

Other important statistics related to the least-squared linear regression is the residual, the residual standard deviation ($s_{y/x}$), the standard deviation of the slope (s_b), and the standard deviation of the intercept (s_a).

$$\text{Residual} = (y - \hat{y}) \qquad (7.20)$$

$$s_{y/x} = \sqrt{\frac{\sum\limits_{i=1}^{n}(y_i - \hat{y})^2}{n-2}} \qquad (7.21)$$

where
   $y_i =$ The observed value.
   $\hat{y}_i =$ The calculated value of y from the regression equation.
   $n =$ The number of pairs of data used in the regression.

$$s_b = \frac{s_{y/x}}{\sqrt{\sum_{i=1}^{n}(x_i - \overline{x})^2}} \tag{7.22}$$

$$s_a = s_{y/x}\sqrt{\frac{\sum_{i=1}^{n}x_i^2}{n\sum_{i=1}^{n}(x_i - \overline{x})^2}} \tag{7.23}$$

Once the regression equation is applied and the values of the slope b, and intercept a, are determined it is important to understand the level of confidence of b, and a, so that accurate conclusions can be drawn from the data. The confidence intervals about the data for the slope b, and the intercept a, are determined by Eq. (7.24) and Eq. (7.25), respectively.

$$b \pm ts_b \tag{7.24}$$

$$a \pm ts_a \tag{7.25}$$

where t = the two-tailed t value at the desired significance level (usually 0.05) with degrees of freedom $v = n - 2$.

The measure of the linear relationship between the variables x, and y, can be determined by the correlation coefficient. The correlation coefficient (r) is calculated by utilizing Eq. (7.26).

$$r = \frac{\sum_{i=1}^{n}\left[(x_i - \overline{x})(y_i - \overline{y})\right]}{\sqrt{\left[\sum_{i=1}^{n}(x_i - \overline{x})^2\right]\left[\sum_{i=1}^{n}(y_i - \overline{y})^2\right]}} \tag{7.26}$$

The value of r will be in the range of ± 1, when the value of $|r|$ is closest to 1 the greater the correlation between the variables exist. The correlation coefficient should not solely be taken as measure of linearity, it should be utilized in conjunction with other information such as the data of the independent variable being evenly distributed with no obvious aberrations and the plot of the residuals should appear to be randomly distributed with no apparent trends. If predictions are to be made from calibration curve produced from the regression equation, the value of $|r|$ should be ≥ 0.9999.

Once the best fit straight line has been determined utilizing the regression equation, it is important to determine the uncertainty associated with the predicted value $\hat{x}$. The predicted value ($\hat{x}$) is calculated by the following equation:

$$\hat{x} = \frac{y_0 - a}{b} \tag{7.27}$$

where
$y_0 =$ The mean of N measurements of y.
The uncertainty for the predicted value $\hat{x}$ is calculated by Eq. (7.28).

$$s_{\hat{x}} = \frac{s_{y/x}}{b} \sqrt{\frac{1}{N} + \frac{1}{n} + \frac{(y_0 - \bar{y})^2}{b^2 \sum_{i=1}^{n} (x_i - \bar{x})^2}} \tag{7.28}$$

The value of $s_{\hat{x}}$ is referred to as the standard error of prediction for $\hat{x}$. The uncertainty of $\hat{x}$ is greatest at extreme ends of the data range and is at a minimum at the points $\bar{x}$, $\bar{y}$. The confidence interval for $\hat{x}$ is denoted by the equation:

$$\hat{x} \pm t s_{\hat{x}} \tag{7.29}$$

The determination of all of the above statistics for the least-squared linear regression is important for interpreting the data. The confidence interval for the slope b is not generally important when the regression line is being used as a calibration curve, since in the line should have a very high correlation for the variables x, and y, and the slope should be significantly different from zero. However, when the regression line is used to determine limits, such as shelf life of a product then the confidence levels of the slope (b) become very important. In this case a wide confidence limit of the slope may cause a short dating of the products shelf life.

It is important to determine the linearity of the calibration line. As previously mentioned the correlation coefficient is not a very good for the determination of linearity, more specific tests for nonlinearity can be utilized. Nonlinearity can be detected by observing the data on a scatter plot, specifically a plot of the residuals. Comparing the residual standard deviation ($s_{y/x}$) with the standard deviation of y (s_y) values from multiple observations for single value of x, and utilizing the F statistic can be a good indicator of linearity or the presence of nonlinearity. The F statistic is calculated from the following equation:

$$F = \frac{s_{y/x}^2}{s_y^2} \tag{7.30}$$

The null hypothesis is $H_0 : s_{y/x} = s_y$ and the alternate hypothesis, $H_1 : s_{y/x} > s_y$. The F test is therefore a one-tailed statistic with $n - 2$ degrees of freedom for $s_{y/x}$, where n is the number of data pairs in the regression data set and $n - 1$ degrees of freedom for s_y

where n is the number of replicate observations for the single value of x. The critical value for F is obtained from the F distribution table for a chosen probability α (usually 0.05), with v_1 degrees of freedom for $s_{y/x}$ and v_2 degrees of freedom for s_y. If the calculated value of F is greater than the critical value found, then the null hypothesis is rejected and the residuals are significantly different than can be attributed to random variation. This result along with a plot of the residuals demonstrating a nonrandom trend would be indicative of nonlinearity. Another approach to evaluating linearity is to determine the fit of the data to a polynomial regression. If the data has a better fit to polynomial regression, it is an indication that there is curvature to the data and there is significant departure from linearity.

Once the calibration line has been determined to have a satisfactory linear fit to the data, it is important to determine if the intercept (a) is significantly different from zero. This can be determined by calculating the confidence interval of the intercept and determine if zero is included in the interval. If it is found that zero is included within the confidence interval, then the intercept is not significantly different from zero (which is the desired result). If it is found that zero is not included in the confidence interval of the intercept, then the intercept is significantly different from zero, which would indicate some bias is influencing the regression line and an investigation of the data should be performed.

7.6 Required Sample Replicates [6]

One of the most common asked questions by analytical chemist is: how many replicate samples do I need for this experiment? The answer is not simple, it depends on a desired limit, the level of confidence required (usually 95 %), and the acceptable uncertainty level of the process which is represented by the standard error of the mean $s(\bar{x}) = \dfrac{s}{\sqrt{n}}$) equation (Eq. (7.4)). The minimum number of replicates can be calculated by the following equation.

$$n_{min} \geq \left[\frac{s}{s(\bar{x})} \right]^2 \tag{7.31}$$

For example, if a method procedure has a standard deviation of 4 % and desires to produce their data with a criterion of ±5 % at 95 % confidence level, what is the minimum number of replicate samples required?

Confidence interval = $t\left(s(\bar{x})\right)$

t=table value for $n-1$ degrees of freedom for the 95 % confidence level

If $n=6$, then $S(\bar{x}) = \dfrac{5}{2.776} = 1.80$

$$n_{min} \geq \left[\frac{4}{1.80} \right]^2 = 4.9$$

This would be raised to five replicates.

7.7 Method Performance [1, 4]

Method validation is the process which is utilized to determine if a method performance is appropriate for its intended use. Method validation consists of evaluating parameters such as accuracy, precision, linearity, limit of quantitation, limit of detection, specificity, and ruggedness.

- Method Precision
  Method precision is defined as closeness among individual measurement for a single sample. Precision is usually determined by the calculation of the standard deviation (Eq. (7.3)) or relative standard deviation (Eq. (7.5)) obtained from replicate measurements of an single sample. Method precision is evaluated as two entities, repeatability and intermediate precision. The conditions under which the measurements are made to determine which type of precision is being estimated.
  Repeatability or also referred to within run precision is performed by a single analyst, on a single instrument, during a single run. This type of precision is an estimate of the variation among replicate measurements in a single run using the same sample throughout the run.
  Intermediate precision is the determination of precision utilizing more variable conditions than used in repeatability. In the determination of intermediate precision variation such as different runs or days, different analysts, different sets of equipment, or any other variables could occur during routine use of the method should be evaluated. It is essential that all the variations applied during the determination of intermediate precision be documented. The variances for repeatability and intermediate precision are shown in Table 7.2.

Table 7.2 Variance for k groups with n replicates per group

Variance	Sum of squares	Degrees of freedom	Mean square	F statistic
Within group (S_w)	$S_w = \sum_{i=1}^{k} \left(x - \overline{X_i}\right)^2$	$N-k$	$MS_w = \dfrac{S_w}{N-k}$	$\dfrac{MS_b}{MS_w}$
Between group (S_b)	$S_b = \sum_{i=1}^{k} \left(\overline{x}_i - \overline{x}_{GM}\right)^2$	$k-1$	$MS_b = \dfrac{S_b}{k-1}$	
Total (S_{tot})	$S_{tot} = S_w + S_b$	$N-1$		

Reprinted from [6] permission Springer Science + Business
K = Number of treatment groups
n = Number of samples per group
N = Total number of samples
$\overline{x}_{GM}$ = Grand Mean for total number of samples (N)
S_w = Sum of squares within group
S_b = Sum of squares between groups
S_{tot} = Total sum of squares
MS_w = Mean square within
MS_b = Mean square between

Utilizing the equations from Table 7.2, Repeatability and Intermediate Precision is calculated as follows:

$$\text{Repeatability}\left(P_r\right) = \sqrt{MS_w} \tag{7.32}$$

An estimation of the Between Group precision is as follows:

$$\text{Between}\left(P_b\right) = \sqrt{\frac{MS_b - MS_w}{n}} \tag{7.33}$$

An estimation of the Intermediate Precision is as follows:

$$\text{Intermediate Precision}\left(P_I\right) = \sqrt{P_r^2 + P_b^2} \tag{7.34}$$

Reproducibility is determined using the same calculations as intermediated precision, except that the variations are between two different laboratories. In many cases reproducibility can be substituted for intermediate precision (see ICH 2Q(R)).

If the precision of the methods need to be compared to determine if the method precision is significantly different from each other, the F statistic can be calculated using the Eq. (7.12).

- Accuracy
  Accuracy is defined as the closeness of a measurement to the true value. Therefore accuracy includes both precision and bias as part of its value. Trueness is usually expressed in terms of bias. Bias can be evaluated when the mean $\left(\bar{x}\right)$ of several measurements is compared to the true value $\left(u_0\right)$. In practice the true value is a certified reference standard spiked in the sample matrix. Bias is therefore calculated as

$$\text{Bias} = \bar{x} - u_0 \tag{7.35}$$

$$\%\text{Bias} = \frac{\bar{x} - u_0}{u_0} \times 100 \tag{7.36}$$

$$\text{Recovery}\left(\%\right) = \frac{\bar{x}}{u_0} \times 100 \tag{7.37}$$

It is usually important to determine if the mean value $\left(\bar{x}\right)$ is significantly different from true value $\left(u_0\right)$. This can be calculated using the student t test at a specified confidence level, usually 95 %, as described in Sect. 7.2. If there is a significant difference between obtained mean value $\left(\bar{x}\right)$ and the true value $\left(u_0\right)$, the measurement is significantly biased and should not be used for accuracy determination.

- Linearity
  The evaluation of linearity is essential for determining if the instrument response is linear with the analyte concentration. The linearity is determined by the evaluation of several concentration levels, generally not less than five, equally spaced over a concentration range utilizing least-square linear regression analysis and the correlation coefficient as discussed in Sect. 7.5. Once linearity is established a working concentration range must be determined to establish that the method is adequate for its intended use. The working range must include the LOQ and level 10 % above the target concentration. The samples used to evaluate the working concentration range must mimic the sample matrix of material which will be analyzed by the method, such as certified reference standard, spiked placebo, or prepared matrix matched standard solutions. If nonlinearity is observed with the matrix sample studies, this may indicate the presence of interfering compounds or other bias present in the method, in any case the method needs to be investigated.

- Limit of Detection (LOD)
  The LOD is the minimum concentration of an analyte that can be detected that is significantly different than the response of the blank. The LOD can be estimated by obtaining the standard deviation of replicate analysis of a blank sample (no analyte in the sample). Generally six to ten replicates taken through the analytical method should be obtained. Using statistics, a limit which only allows 5 % of the distribution to be considered a false positive result can be calculated by multiplying the standard deviation by a factor based on a one-tailed student t value, with infinite degrees of freedom, and adding that to the mean value x_0 of a blank. Thus by choosing α at 0.05 the t value is 1.65 and the limit value is $x_0 + 1.65\sigma_0$. Therefore any value equal to or greater than the limit value is considered a positive finding and the value is statistically above zero. However, we also have to consider values that appear to be negative, but may in fact be positive (false negative). Using the same statistical approach, a concentration can be determined which the limit value will cut off an area β of the expected distribution for the calculated concentration. This new, higher, calculated concentration is called the *Limit of Detection*. Generally the significance level for α and β is the same, 0.05, allowing for a 5 % false negative rate. The limit of detection for a method where no correction for baseline is performed can be shown in the following equation:

$$LOD = x_0 + \sigma t_{\alpha,v} + \sigma t_{\beta,v} \tag{7.38}$$

$$LOD = x_0 + 1.65\sigma + 1.65\sigma = x_0 + 3.3\sigma$$
$$\text{If baseline corrected, } x_0 = 0 \tag{7.39}$$
$$\text{then LOD} = 3.3\sigma$$

where
x_0 = Mean of blank measurements
σ = Standard deviation of the blank
t = the value of the one-tailed student t with infinite degrees of freedom

- To compare the calculated LOD from detectors with different responses, the LOD can be normalized by dividing the LOD with the slope of the calibration line. The more general equation for LOD is as follows:

$$LOD = \frac{3.3\sigma}{S} \tag{7.40}$$

where
S = Slope of the analyte calibration line

- The equation for LOD is consistent with the equation for LOD within the ICHQ2(R) guideline.

- Limit of Quantitation (LOQ)
  The LOQ is the lowest level of an analyte that can be quantitated within defined confidence level. Similar to approach for the LOD, the standard deviation of multiple replicates of the blank, σ, is utilized in the calculation. A value of 10σ is frequently used as an acceptable level. Also to normalize LOQ among detectors of varying responses, the LOQ is divided by the slope of the analyte calibration line, thus resulting in a general equation for LOQ as follows:

$$LOQ = \frac{10\sigma}{S} \tag{7.41}$$

7.8 Uncertainty Measurement [4, 5]

Measurement is a process, in which specified procedures are performed to determine a value. In a measurement process even when all the measurement factors are controlled, repeated observation using the same process under the same condition are rarely found to be identical. This is due to the variables such as operator, reference standards, materials, instrument, environment, calibration, and test methods. Therefore measurement results are never the true value because of the uncertainty associated with them. The following list contains the need for the determination of uncertainty

1. The customer needs to know that measurement uncertainty has to be taken into account particularly when regarding the conclusions being drawn from the data.
2. Testing laboratories shall have procedures for estimating uncertainty of measurement.
3. The uncertainty values may be required in the report of analysis
4. In calibration, uncertainties have to be stated in the certificate of analysis so that it can be utilized by the user of the equipment.

There are several essentials of measurement uncertainties that must be considered such as:

- Data Systems
- Environment
- Human Error
- Methodology
- Physical Properties
- Reference Standard
- Statistics

There are two methods of estimating uncertainty. Type A, where the estimation is based on statistical analysis of replicate measurements. Type B, estimations come from any other sources such as uncertainty value in reference standard, value in a calibration certificate, specification on a volumetric flask or balance. Type A and Type B estimations are combined for the measurement of uncertainty. Individual uncertainty is known as a *standard uncertainty* (u) and a reported uncertainty is known as an *expanded uncertainty* (U).

There are two rules for calculating uncertainties in the analytical result.

1. If the quantities are added or subtracted to obtain a result y, the uncertainty in y is the uncertainty $u(x_i)$ in x_i
2. If the quantities are multiplied or divided, the expression for y the contribution of uncertainty in y, $u(y)/y$ is the uncertainty $u(x_i)/x_i$ in x_i.

In more general cases, where the result is due to several different algebraic operations the uncertainty must be obtained by a different approach. In this case the approach is to determine the change in the result, y based on the uncertainty in, x. That is the uncertainty change in x_i times the rate of change in, y with x_i. This results in the uncertainty calculation for $u_i(y)$ in y, with an uncertainty $u(x_i)$ in x_i. This is shown below in Eq. (7.42).

$$u_i\left(y\right)=c_i u\left(x_i\right) \tag{7.42}$$

where
c_i=The slope of the line of y against x_i

Utilizing Eq. (7.42) for the combination of uncertainties, when the contributions are independent of each other, they combine as the root sum of their squares. This is shown below in Eq. (7.43)

$$u\left(y\right)=\sqrt{\sum_{i=1}^{n}c_i^2 u\left(x_i\right)^2} \tag{7.43}$$

Table 7.3 below describes the combination of uncertainties for independent contributions.

Calculation of result, y	Uncertainty u_y in y from x_i
$y = x_1 + x_2$	$u_y = \sqrt{u_{x_1}^2 + u_{x_2}^2}$
$y = x_1 - x_2$	$u_y = \sqrt{u_{x_1}^2 - u_{x_2}^2}$
$y = x_1 \times x_2$	$u_y y = \sqrt{\left(u_{x_1} x_1\right)^2 + \left(u_{x_2} x_2\right)^2}$
$y = x_1/x_2$	$\dfrac{u_y}{y} = \sqrt{\left(\dfrac{u_{x_1}}{x_1}\right)^2 + \left(\dfrac{u_{x_2}}{x_2}\right)^2}$

Table 7.3 Combination of uncertainties

Reprinted from [6] permission Springer Science + Business

Combining uncertainties from contributions which are not independent from each other requires a much more complex relationship, which can be found in [6]. Reporting the measurement uncertainty is usually in the form as an expanded uncertainty, U. The expanded uncertainty, U is determined by multiplying the standard uncertainty, u by a factor k. The determination of factor k is based on the student t value at a desired confidence level (usually 95 %). For a confidence of 95 % (0.05), the t value at infinite degrees of freedom is 1.96 and the k factor used is rounded to 2. Therefore the reported uncertainty U will be two times the standard uncertainty u. The resulting value, U is commonly rounded up to one or two significant figures.

7.9 Sampling Strategies [6, 4]

Sampling is one of the most important processes to insure that the data obtained from the analysis is appropriate for its intended use. Improper sampling can contribute the greatest amount of uncertainty to the results. Therefore the determination of a sampling process which is suitable for the method must be developed and executed. From a statistical aspect sampling is a subset of items taken from a larger population which allows a minimum bias and is representative of that population. There are many different types of samples, some of which are described below:

Bulk Sample—The primary material that requires analysis such as the lot or batch of an evidence material

Subsample—A selected portion of a sample. This can be the removal of selective samples from the bulk, or the removal of samples from a laboratory sample for analysis.

Composite Sample—The combining of a collection of subsamples into a uniform homogenized sample for analysis. When applicable, it serves as a reduction in use of analytical resources.

There are several strategies to sampling that must be considered so that the sample analysis is adequate for its intended use. It is important that each sample taken has an equal probability of containing all the properties of the bulk. Randomization of sampling is a strategy which is usually used as the best approach to minimize unwanted biases. Although randomization can minimize the bias among the sub-samples, it does not ensure that the samples are representative of the bulk material. Therefore it is necessary to obtain samples that are unbiased but also representative of the bulk. There are several sampling strategies which have different variances. It is the goal to select those strategies which deliver small variances along with randomization procedures which provide unbiased results.

• Simple Random Sampling
  Simple random sampling provides the probability that every subsample chosen has an equal chance of containing all of the properties found in the bulk sample. In the case of choosing items from a number of discrete items such as drums, packages, and bottles random number generators are used to choose the items in a random order. For particulate matter such as powder, the technique of repeated quartering is utilized, but care has to be taken not to increase the bias due to segregation of particle size or other physical properties which are observed. Simple random sampling is the easiest to implement, however because of the variation between items chosen it would be a poor choice for sampling material which nonhomogeneous, in this situation simple random sampling would have the highest variation of the sampling strategies.
  Statistical treatment for simple random sampling is dependent on the number of sample, n taken from the sample population, N. The standard error of the mean $s(\bar{x})$ due to sampling is given by Eq. (7.45)

$$S(\bar{x}) = s_{sam} \sqrt{\frac{1-f}{n}} \qquad (7.44)$$

where
s_{sam} = The standard deviation of the sampling process
$f = n/N$

• The standard deviation of the sampling process, s_{sam} can be obtained from the between group component of variance described in Table 7.2 (between group). When N is very large or the sampling n is less than 10 % of the population N, then f can be ignored and Eq. (7.45) is reduced to Eq. (7.45).

$$S(\bar{x}) = \frac{s_{sam}}{\sqrt{n}} \qquad (7.45)$$

• When dealing with particulate matter, the sampling statistics can also be treated as a large N situation and Eq. (7.45) can be utilized.
• If it is determined that analytical uncertainty is very small, then s_{sam} is the standard deviation (s) of the observation x_i.

- Stratified random sampling
  In the stratified random sampling process the sampled population is divided into segments and each segment is sampled as per the simple random sampling process. The number of items selected from each segment is dependent on the intended use of the measurement. There are several considerations that must be addressed when applying the stratified random sampling process.

  - There must be equal number of items per segment and the same number of items is chosen from each segment. Having each segment of equal size allows the use of composite sampling of the whole without introducing a biased estimate of the bulk composition
  - Proportional sampling is the process of taking the number of items from each segment proportional to the fraction of each segment in the bulk. That is if the bulk has samples that are different, then random samples from each segment would be chosen such that the total samples taken are in proportion with each of the types of samples in the bulk. Proportional sampling also provides the opportunity for composite sampling. The variance observed from proportional sampling is smaller than that produced from simple random sampling, especially when there is no information available about the variances of the individual segments.
  - The determination of the number of samples per segment is based on the size and the standard deviation of each segment. This strategy provides the smallest variance for the total number of samples in the segment. The optimum number samples (n_i^{opt}) to take from the ith segment of n samples are determined by Eq. (7.46).

$$n_i^{\text{opt}} = \frac{P_i \sigma_i}{\sum P_i \sigma_i} n \qquad (7.46)$$

where
P_i = The proportion of segment i in the population
σ_i = The standard deviation of samples in the ith segment

- The statistical treatment for stratified random sampling requires the determination of the mean $\bar{x}_i$, total mass in each segment m_i, the standard deviation $s_{\text{sam},i}$, and the number samples in the segments n_i. The proportion, P_i of each segment in the bulk is shown below.

$$P_i = \frac{m_i}{\sum_{i=1}^{n_i} m_i} \qquad (7.47)$$

$$\text{The mean of the Bulk } \bar{X} = \sum_{i=1}^{n_i} P_i s_{\text{sam},i} \qquad (7.48)$$

$$\text{The variance} S^2\left(\bar{X}\right) = \sum_{i=1}^{n_i} \frac{P_i^2 s_i^2 \left(1 - f_i\right)}{n_i} \qquad (7.49)$$

$$\text{The standard error of the mean} \, s\left(\overline{X}\right) = \sqrt{s^2\left(\overline{X}\right)} \qquad (7.50)$$

$$\text{The total mass of analyte} = m_{\text{tot}}\overline{X} \qquad (7.51)$$

$$\text{Standard uncertainty from sampling} = m_{\text{tot}}s\left(\overline{X}\right) \qquad (7.52)$$

- Example:
  As part of a drug arrest the police confiscated 135 bags of white powder of alleged heroin containing 200 g per bag and 280 bags of white powder of alleged heroin containing 5 g per bag which has been cut with diluents. Each of the 135 bags is found to be similar in composition as is each of the 280 bags. Ten bags are sampled from each population. An estimate of the total amount of heroin is required along with measurement uncertainty. The testing is shown in Table 7.4

Table 7.4 Stratified random sampling strategy

	Large bags	Small bags	Total
Analytical results			
Mass, m_i	27,000 g	1,400	
Total mass, m_{tot}			28,400 g
Total items, N_i	135	280	
Test sample number, n_i	10	10	
$\overline{X}_i$ of heroin g/100 g	50	17	
S_i g/100 g	4.0	2.5	
Calculations			
$P_i = \dfrac{m_i}{\sum\limits_{i=1}^{n_i} m_i}$	0.95	0.049	
$P_i\overline{X}_i$	47.5	0.833	
$f = n_i/N_i$	0.074	0.036	
$s^2\left(\overline{X}\right) = \sum\limits_{i=1}^{n_i} \dfrac{P_i^2 s_i^2\left(1-f_i\right)}{n_i}$			
Mean concentration $\bar{x}_i$	1.34	0.0015	
$s\left(\overline{X}\right)$			
Total heroin $m_{\text{tot}}\overline{X}/100\,g$			(47.5+0.833)=48.3
Uncertainty $m_{\text{tot}}s\left(\overline{X}\right)/100\,g$			$\sqrt{1.34+0.0015} = 1.16$
			28,400×48.3/100=13,717 g
			13.72 kg
			28,400×1.16/100=329.4 g
			% UC=329.4/28,400=1.15 %

References

1. Freund JE (1967) Modern elementary statistics, 3rd edn. Prentice Hall, Upper Saddle River
2. Draper N, Smith H (1981) Applied regression analysis, 2nd edn. Wiley, New York
3. Box GEP, Hunter WG, Hunter SJ (1978) Statistics for experimenters, an introduction to design, data analysis, and model building. Wiley, New York
4. Ellison SLR et al (2009) Practical statistics for the analytical scientist, 2nd edn. RSC, Cambridge
5. ISO Guide 98 (1995) Guide to the expression of uncertainty in measurements. International Organization for Standardization, Geneva
6. Catalano T (2013) Essential elements for a GMP analytical chemistry department. Springer, New York

Chapter 8
Conclusion

Over the last two decades, forensic science received very high notability in the public eye. Mainly because of the popularity of television shows such as CSI Miami and CSI New York, to name a few, put a twist on standard police procedure. In this new world of policing, crimes are solved using high tech scientific technologies, very rapidly and with 100 % certainty. However, there are great differences in the practice of forensic science across various jurisdictions, many due to funding, equipment, and the availability of skilled and well-trained personnel. This exists because many of the operational principles and procedures in the forensic sciences disciplines are not standardized. Generally there are no standard protocols directing the practices for a given discipline, other than DNA. Therefore the quality of practices, in most disciplines, varies greatly. Because the forensic data is generally used in a court of law (criminal and civil), it is critical to determine whether the forensic data can be accepted as evidence. Although disciplines such as forensic chemistry and toxicology are derived from the Scientific Method approach, they do not require a standardized approach to applying the technology in a manner which would be acceptable in the scientific community which developed and utilizes the technology.

In order to improve the validity of forensic evidence presented in court, it became part of our law that the judge would act as the gatekeeper for determining which scientific forensic evidence are appropriate for consideration to be presented in court. Two court rulings were passed over the years to set some criteria for determining whether the scientific forensic evidence was suitable for presentation in court. Frye vs. United States, Court of Appeals of District of Columbia, in 1923, proposed that the scientific approach needed to be sufficiently established so that it had gained general acceptance in the relevant scientific community and had to be relevant to issues being considered.

In 1993 Jason Daubert and Eric Schuller were two children who claimed that they were born with serious defects based on their mother's use of Benedectin during her pregnancy. They sued Merrell Dow the manufacture of the drug. The court ruling from the Daubert decision proposed that scientific evidence be derived from

© Thomas Catalano 2014
T. Catalano, *Good Laboratory Practices for Forensic Chemistry*,
SpringerBriefs in Pharmaceutical Science & Drug Development,
DOI 10.1007/978-3-319-09725-1_8

methodology that can be scientifically supported and contain associated error measurements along with the determination of confidence intervals where appropriate. Although the rules like Frye and Daubert have been identified, they have not resulted in any meaningful limitations on the admissibility of forensic evidence and as a result are not practiced in many state and local jurisdictions. The courts still rely on precedent and every non-scientific based decision becomes a precedent binding on future cases. Recently other groups have immerged to address the problems associated with validity the forensic evidence presented in court. One of these groups is the Scientific Working Group (SWG) for various forensic disciplines. The group addresses many of the issues detailed in this work, but without the emphasis on the implementation. The American Society of Crime Laboratory Directors (ASCLD) is another organization working on setting standards for forensic laboratories. Their major concern is the accreditation of the forensic laboratories; however, they are also working towards setting standards.

In January 2014, as a result of the National Academy of Sciences (NAS) published report of February 18, 2009, the National Institute of Standards and Technology (NIST) and the Department of Justice (DOJ) created the National Commission on Forensic Science (NCFS) that will work to improve the practice of forensic science by developing guidances and policy recommendations for the US Attorney General. Under this administration, a number of interdisciplinary working groups have been launched to produce technical publications and other forms of critical guidances for the forensic science community. The areas being addressed are Forensic Research, Development of Standards, Guidelines and Best Practices, Scientific Capacity, New Technology and Tools, Workshops and symposia, Education and Training, and International Collaborations.

In this book the emphasis was placed on the many guidances within the pharmaceutical industry which describe in detail the standards, protocols, best practices, training, and requirements for the acceptance of data for submission to regulatory agencies. These activities, which were successfully utilized for many years, were discussed in detail with the intention to be applied to forensic chemistry, without changing the conceptual rigor intended in the document. The documents described within guidances such as Good Manufacturing Practices (GMP), Good Laboratory Practices (GLP), International Committee on Harmonization (ICH), Quality Assurance (QA), and Quality Risk Management were reviewed and considered for application to forensic chemistry. Work processes such as method development, validation, training, reference standards, technology transfers, reports of analysis, statistics, and sampling are detailed in the book for application to forensic chemistry. The ISO documents are also reviewed. Each relevant clause was discussed to ensure their understanding towards their implementation. Having described much of the guidances in enough detail for their implementation is the first step in making the enforcement capable and desirable. It is the driving for standardization from the bottom up that will eventually make it become a practice. I believe the discussions presented will allow a sufficient understanding of the concepts and their implementation to motivate the forensic scientists and their management to view these issues as an absolute problem that must be fixed in order for forensic chemistry to become well respected within the forensic community similar to DNA testing.

Index

A
Accuracy, 8, 11, 19, 34–36, 57, 58
The American Society of Crime Laboratory
Directors (ASCLD), 9, 68

B
Between variance, 58, 63

C
Conclusions, 3–6, 8, 9, 29, 51, 54, 60, 67–68
Conduct of Laboratory Study, 28, 31
Confidence intervals, 9, 20, 35, 45, 50, 54–56, 68
Correlation coefficient, 34, 54, 55, 59
Current state, 7–8

D
Daubert, J., 8, 9, 67, 68
Degrees of freedom (DOF,ν), 45–47, 49, 50,
54–57, 59, 60, 62
Department of Justice (DOJ), 9, 68
Diode array, 26, 27, 34
DNA, 7, 8, 67, 68

E
Expanded uncertainty, 61, 62
Experimentation, 3–5

F
Forensic chemistry, 1, 2, 7–9, 11, 14, 67, 68
F-test, 47, 49, 50
Frye, 8, 9, 69

G
Gas chromatography (GC), 18, 25
Good Laboratory Practices (GLP), 2, 28–31,
33, 68
Good Manufacturing Practices (cGMP),
2, 11–28, 33, 68
(21CFR) Guidance's, 11–31

H
Handling test article, 28, 30
High performance liquid chromatography
(HPLC), 13–18, 26
How to improve, 8–9
Hypothesis, 4–6, 47, 48, 55, 56

I
Intermediate precision, 34, 35, 57, 58
International Committee on Harmonization
(ICH), 2, 33–39, 68
Interpretation, 6, 7, 17, 38
IQ/OQ/PQ, 25–27
ISO 17025, 41

© Thomas Catalano 2014
T. Catalano, *Good Laboratory Practices for Forensic Chemistry*,
SpringerBriefs in Pharmaceutical Science & Drug Development,
DOI 10.1007/978-3-319-09725-1